Particle Technology Forum

Held at the 2023 AIChE Annual Meeting

Orlando, Florida, USA
5-10 November 2023

ISBN: 978-1-7138-9317-2

Printed from e-media with permission by:

Curran Associates, Inc.
57 Morehouse Lane
Red Hook, NY 12571

Some format issues inherent in the e-media version may also appear in this print version.

Printed with permission by Curran Associates, Inc. (2024)

For permission requests, please contact AIChE
at the address below.

AIChE
120 Wall Street, FL 23
New York, NY 10005-4020

Phone: (800) 242-4363
Fax: (203) 775-5177

www.aiche.org

Additional copies of this publication are available from:

Curran Associates, Inc.
57 Morehouse Lane
Red Hook, NY 12571 USA
Phone: 845-758-0400
Fax: 845-758-2633
Email: curran@proceedings.com
Web: www.proceedings.com

Particle Technology Forum

Held at the 2023 AIChE Annual Meeting

Orlando, Florida, USA
5-10 November 2023

AIChE
The Global Home of Chemical Engineers

TABLE OF CONTENTS

CRYSTALLIZATION IN PROCESS DEVELOPMENT

PARTICLE CHARACTERIZATION

FLUIDIZATION: EXPERIMENTAL INVESTIGATION AND MODELING OF FLUIDIZATION PROCESSES

FLUIDIZATION: FUNDAMENTALS

FLUIDIZATION: MODELING

FUNCTIONAL NANOPARTICLES AND NANOCOMPOSITES

NANOSTRUCTURED MATERIALS FOR PHARMACEUTICAL APPLICATIONS

NOVEL NANOPARTICLES AND NANOSTRUCTURED CATALYSIS FOR ENERGY AND ENVIRONMENTAL APPLICATIONS

NOVEL NANOPARTICLES AND NANOSTRUCTURED MATERIALS FOR PHARMACEUTICALS AND MEDICAL APPLICATIONS

PARTICLE FORMATION PROCESSES

PARTICLE BREAKAGE

PARTICLE TECHNOLOGY IN PRODUCT DESIGN AND MANUFACTURING

PARTICULATE MIXING AND SEGREGATION

PARTICULATE SYSTEMS: DYNAMICS AND MODELING: APPLICATIONS

PARTICULATE SYSTEMS: DYNAMICS AND MODELING: DISCRETE/CONTINUUM MODELS

PARTICULATE SYSTEMS: SOLIDS HANDLING, PROCESSING, CONVEYING, SEPARATION, AND HEAT TRANSFER

PHARMACEUTICAL POWDER AND PARTICULATE SYSTEMS

POSTER SESSION: PARTICLE TECHNOLOGY FORUM

POWDER AND PARTICULATE CHARACTERIZATION AND MEASUREMENT

PARTICULATE PROCESS MODELING AND PRODUCT DESIGN

PARTICLE ENGINEERING APPLICATIONS: ADDITIVE MANUFACTURING, ENERGY STORAGE AND CARBON CAPTURE

POWDER HANDLING AND PROCESSING FOR BATTERY MATERIALS

HONORARY SESSION: RAY COCCO'S SPECIAL TOPICS

INVITED: CELEBRATING 30 YEARS OF THE PTF (THE FIRST 15 YEARS)

INVITED: CELEBRATING 30 YEARS OF THE PTF (LAST 15 YEARS)

92a Development of Automated Milling Platform for Crystal Engineering

Shailesh Agrawal, Saurin Hiren Rawal, Jeremy Merritt

Development of automated milling platform for crystal engineering

Shailesh Agrawal, Saurin H Rawal, Jeremy Merritt*

**Presenting Author*

Synthetic Molecule Design and Development, Eli Lilly & Company.

Wet milling is widely used in pharmaceutical industry as a leverage to modify both; size and morphology of the final crystalline drug substance (DS) as these play a pivotal role in formulation of an appropriate drug product [1]. A common approach to modify crystal size and morphology is to use multiple thermal cycles in conjunction with milling, that involves cyclical dissolution and recrystallization steps[2]. There is a widespread adoption of autonomous reactor operations coupled with process analytical technology that helps in accelerating crystallization development but there aren't significant inroads in extending robust automation to slurry milling. Herein, we demonstrate an autonomous milling setup paired with an automated crystallizer setup that can extend equipment utilization in an industrial setup and help develop data rich milling processes in a fashion that would aid current aggressive drug substance/ drug product development timelines.

We will discuss the development and deployment of a customized automated platform to perform milling in conjunction with thermal cycling to achieve desired improvements in the DS physical properties. Lovastatin, a commercially available DS and notorious for yielding needle shaped crystals [3] is used as a model compound to demonstrate the capabilities of the platform. We will use this compound to demonstrate the time to develop the process and corresponding improvement in physical properties.

The study will briefly discuss the efforts towards building a high-fidelity mathematical model for the combined milling and thermal cycling process. It is envisioned that coupled with a digital twin[4], the automated platform will significantly accelerate the development of an optimized milling process to achieve desired size and morphology.

References

1. Salvatori, F. and M. Mazzotti, *Manipulation of Particle Morphology by Crystallization, Milling, and Heating Cycles—A Mathematical Modeling Approach.* Industrial & Engineering Chemistry Research, 2017. **56**(32): p. 9188-9201.
2. Wilson, D., et al., *Particle engineering of needle shaped crystals by wet milling and temperature cycling: Optimisation for roller compaction.* Powder Technology, 2018. **339**: p. 641-650.
3. McGinty, J., et al., *Effect of Process Conditions on Particle Size and Shape in Continuous Antisolvent Crystallisation of Lovastatin.* 2020. **10**(10): p. 925.
4. Szilágyi, B., et al., *Digital Design of the Crystallization of an Active Pharmaceutical Ingredient Using a Population Balance Model with a Novel Size Dependent Growth Rate Expression. From Development of a Digital Twin to In Silico Optimization and Experimental Validation.* Crystal Growth & Design, 2022. **22**(1): p. 497-512.

92b Process Control and Design of the Continuous Crystallization of a Polymorphic Agrochemical

Montgomery Smith, Paul Larsen, Venkat Bhamidi, Zoltan Nagy

Crystallization is an essential process of solids manufacturing and is left inadequately designed in several fields, including agrochemical manufacturing. Inadequately designed crystallization protocols can lead to particles with undesired physical or chemical characteristics, such as particle morphology, polymorphism, crystal size distribution/aspect ratio, manufacturability, and overall crystal quality.

In this work, we demonstrate the multi-objective process control and design of an agrochemical crystallization by i) controlling the produced polymorphic form via combined cooling and antisolvent crystallization, ii) controlling the final crystal size distribution (CSD) and resultant manufacturability with wet milling, and iii) designing the continuous crystallization for industrial manufacturing of the model system. Variations in combined cooling and antisolvent crystallization operating trajectory can have a dramatic impact on the generated polymorphic form (Kshirsagar *et al.*, 2023). A polymorphic form design space was generated by a data-rich design of experiments (DoE) enabled by the inclusion of *in-situ* process analytical technology (PAT) tools. This design space allowed for informed crystallization operating trajectory design. Following the isolation of the singular polymorphic forms, concerns about manufacturability arose from the presence of a high aspect ratio (AR) morphology. High AR crystals are challenging to industrial processes in terms of poor mixing and filtration time, the inclusion of wet milling with specific crystallization operating trajectories improved both product CSD and AR regarding manufacturability (Eren *et al.*, 2021). Using the polymorphic design space, wet milling, and targeted antisolvent and cooling crystallization operating trajectories, manufacturing in the presence of high AR morphology was greatly improved. Further applying this process control, design, and system information to the continuous crystallization of the model system reduces manufacturing time and variability during industrial manufacturing.

References

1. Kshirsagar, S., Szilagyi, B., Nagy, Z.K., 2023. Experimental Design for the Efficient Determination of the Crystallization Kinetics of a Polymorphic System in Combined Cooling and Antisolvent Crystallization. Crystal Growth & Design.. doi:10.1021/acs.cgd.2c01075.
2. Eren, A., Szilagyi, B., Quon, J.L., Papageorgiou, C.D., Nagy, Z.K., 2021. Experimental Investigation of an Integrated Crystallization and Wet-Milling System with Temperature Cycling to Control the Size and Aspect Ratio of Needle-Shaped Pharmaceutical Crystals. Crystal Growth & Design 21, 3981–3993.. doi:10.1021/acs.cgd.1c00308.

92c Recovery of Valuable Metals from Industrial Leachates of Spent Ni Metal Hydride Batteries By Precipitation Processes.

Boris Guzhov, Laurent Cassayre, Nicolas Coppey, Béatrice Biscans

The recycling of spent nickel metal hydride (Ni-MH) batteries from hybrid electric vehicles serves as an increasingly-attractive secondary source of critical materials thanks to the recent development of hydrometallurgical processes.[1,2] Ni-MH batteries contain multitude of "critical" and "strategic" metals: La, Ce, Nd, Pr, Co, Mn, Ni, etc. and constitute a significant potential source of secondary raw materials with the marketing of approximately one billion cells every year for the past 15 years. Hydrometallurgical process consists in attacking the powders resulting from the grinding of batteries, by an acid (leaching step) to dissolve the elements in order to recover them by subsequent separation operations, such as precipitation.

One possible flowsheet of selective separation of the critical metals by successive precipitation is proposed in Figure 1. The first step after leaching, often reported in the literature, is the recovery of rare earths elements (REEs) by precipitation from pregnant leach solutions (PLS) in sulfate media, using Na_2SO_4 [3]. However, little consideration is given as to whether and how sodium ions influence the precipitation efficiency and selectivity, and detailed phase characterization of the products is rarely reported. After REEs separation the obtained filtrate contains a high amount of Ni that can be recovered by specific precipitation. Nevertheless, before producing Ni components, it is necessary to withdraw impurities such as Fe and Al contained in the PLS by selected preliminary precipitation processes. Very few routes of Fe/Al elimination from the real leachates were explored in the literature. We propose to integrate this step in the recycling process. Finally, several possibilities of Ni precipitation from the final purified leachate can be envisaged.

This work focuses on a better understanding of the precipitation phenomenon and mass balances in the industrial PLS complex system and for the three successive precipitation steps (REEs, Fe/Al, Ni), by coupling pilot-scale experiments on industrially-sourced PLS, in-depth precipitated particles identification, and thermodynamic calculations.

The initial PLS used in this study was provided by SNAM company and originates from the pilot-scale leaching of industrial samples of spent Ni-MH battery powders in H_2SO_4 media, according to a process described in [2]. The battery waste leachate contains 46 $g.L^{-1}$ Ni, 13 $g.L^{-1}$ REEs and 11 $g.L^{-1}$ of other metals like Co, Mn, Al and Fe.

We have applied a selective precipitation pathway to recover REEs and Ni from such a complex solution, without entrapping impurities such as Fe and Al. In our study, for each precipitation experiment, 1.5 L of PLS was added to a 2L double-jacketed agitated batch reactor (550rpm) and heated at 60°C; solution temperature was regulated and pH was monitored. In a standard experiment, a 2.9 M Na_2SO_4 solution was maintained at 40 °C and added to the PLS at a constant flow rate. Suspension volumes were regularly sampled, filtered, and solutions were analyzed by ICP-OES. The precipitation efficiency was calculated as the proportion of the REEs that have precipitated to the initially present in the solution. At the end of Na_2SO_4 addition, the suspension was kept stirred at constant temperature for 1 h and was then filtered to recover a Ni-rich solution on the one hand and REEs-rich filtered cake on the other hand. The obtained REEs crystals were washed and dried before in-depth characterization using multiple techniques: SEM-EDX, powder XRD, TGA, microwave digestion in aqua regia and ICP-OES.

An experimental parametric study on temperature, Na:REE ratio and Na_2SO_4 flowrate, allowed to obtain suited chemical conditions (Na:REEs = 3.6 and T = 60 °C), for high (>95%) REEs precipitation yields. We showed that the precipitation was selective i.e. only REEs precipitate while Ni and other elements remain in solution.

In parallel, the variation of REEs precipitation yield from the PLS via Na_2SO4 addition was evaluated by thermodynamic modeling using the OLI studio module (v11.0) of the OLI.Systems software which includes an accurate thermodynamic description of rare earth sulfates recently developed by Anderko et al.[4]. This program implements the Mixed-Solvent Electrolyte (MSE) model for concentrated electrolyte solutions. The OLI-MSE database contains a specific thermodynamic description of the system rare earths sulfates/sodium sulfate/sulfuric acid /water and thus is well adapted to represent the chemical system.

Calculations were carried out with the various initial PLS conditions (pH, temperature, and ion concentrations). They foresee the precipitation of the compounds $NaLa(SO_4)_2.H_2O$, $NaCe(SO_4)_2.H_2O$, $NaNd(SO_4)_2.H_2O$ and $NaPr(SO_4)_2.H_2O$ in all conditions, which is in good agreement with XRD characterizations. The calculated REEs aqueous concentrations were also close to experimental measurements. At 60 °C, calculations show that the solubilities of all lanthanide-alkali double sulfates decrease with increasing Na^+ concentration, which explains why increasing the Na:REEs molar ratio improves precipitation efficiencies at constant pH. Moreover, for a Na:REEs molar ratio of 4:1, increasing the temperature decreases the double sulfates solubilities (apart from $NaCe(SO_4)_2.H_2O$ which shows the opposite trend).

The XRD analysis of the product as well as of synthetic double sulfate salts allowed us to prove that the obtained product was a solid solution rather than a mixture of individual salts as it was supposed in previous works. The chemical formula of the obtained phase was $(Na_{0.9}K_{0.1})$ $(La_{0.65}Ce_{0.24}Pr_{0.04}Nd_{0.07})$ $(SO_4)_2{\cdot}H_2O$, where the relative proportion of REEs was the same as in the initial PLS.

The second step of the treatment is the withdrawal of Fe and Al impurities, leaving the valuable elements Ni, Co, Mn in the PLS. The composition of the PLS after REEs precipitation is given in Table 1. The potential of the initial solution was measured and was about 0.64 V vs. SHE at 60 °C. Knowing that the potential of the solutions is driven by the couple Fe^{3+}/Fe^{2+} in this system, and taking the Nernst equation, the ratio Fe^{2+}/Fe^{3+} can be calculated and equals 20, which corresponds to 5% of $Fe^{3+}$from Fe_{total}. Meanwhile, Fe^{3+} precipitates at $pH < 3$, while Fe^{2+} can be completely removed from the solution by precipitation only at pH around 8. The latter will affect the quality of the Ni-contain product, which can precipitate also in the same pH range. Hence, we propose to use H_2O_2 as an oxidizing agent in order to transform the portion of Fe^{2+} existing in the solution in Fe^{3+}before precipitation with NaOH solution, allowing the removal of all iron at low pH. Then, pH was increased until pH=5 for precipitation of Al^{3+}. The results show that accurate pH choice is a key factor for the selective recovery of Fe and Al from the PLS. The enhanced precipitation of Fe (>99%) was obtained via the oxidation of Fe^{2+} to Fe^{3+} using H_2O_2 solution. On the other hand, the obtained results also highlight the complete Al removal avoiding Ni losses. Thermodynamic calculation predicted precipitation of 99.9% of Al at pH 4.9.

The final step is to find a method for recovering Ni from the purified PLS. Several methods were investigated, leading to hydroxide or oxalate precipitation according to the selected reactant. The same precipitation efficiencies were reached using both methods: 98.1% of Ni precipitated at pH 9.5 as a hydroxide, while 99.6% of Ni precipitated at pH 0.2 in oxalate form. Nevertheless, in the case of hydroxide precipitation, an amorphous product with poor filterability (residual moisture of 77 wt%) was obtained which affected the washing and resulted in a high Na content in the solid phase. These observations were confirmed by XRD analysis.

In contrast, the precipitated oxalate had a well-crystallized structure and XRD analysis detected only one major phase, which corresponds to nickel oxalate dihydrate. This was confirmed by TGA analysis. The absence of the sodium sulfate on the diffractogram confirms the low Na content, determined by ICP-OES analysis

The study of the final stage of leach solution treatment showed the possibility of complete Ni, Co and Mn removal using either 4 M NaOH or 3.8 M H C O solutions. However, the obtained hydroxide had a high crystal structure disorder, regardless of the use of the seeding technique to prevent the formation of α–$Ni(OH)_2$ polymorph, which led to a poor washing. As a result, the solid phase had high sodium content due to the retaining of the mother solution concentrated in Na^+ and SO_4^{2-}, and the particles were highly agglomerated. The precipitated oxalate $NiC_2O_42H_2O$, in contrast, had a well-crystalized structure that allowed to eliminate the undesirable impurities from the solid phase.

[1] L. Cassayre, B. Guzhov, M. Zielinski, B. Biscans, Chemical processes for the recovery of valuable metals from spent Nickel Metal Hydride batteries: a review. Renewable and Sustainable Energy Reviews 170 (2022) 112983. doi:10.1016/j.rser.2022.112983

[2] M. Zielinski, L. Cassayre, P. Destrac, N. Coppey, G. Garin, B. Biscans, Leaching mechanisms of industrial powders of spent nickel metal hydride batteries in a pilot-scale reactor, ChemSusChem. 4 (2020) 616–628. doi:10.1002/cssc.201902640.

[3] A. Porvali, B.P. Wilson, M. Lundström, Lanthanide-alkali double sulfate precipitation from strong sulfuric acid NiMH battery waste leachate, Waste Manag. (2017). doi:10.1016/j.wasman.2017.10.031.

[4] G. Das, M.M. Lencka, A. Eslamimanesh, P. Wang, A. Anderko, R.E. Riman, A. Navrotsky, Rare earth sulfates in aqueous systems : Thermodynamic modeling of binary and multicomponent systems over wide concentration and temperature ranges, J. Chem. Thermodyn. 131 (2019) 49–79. doi:10.1016/j.jct.2018.10.020.

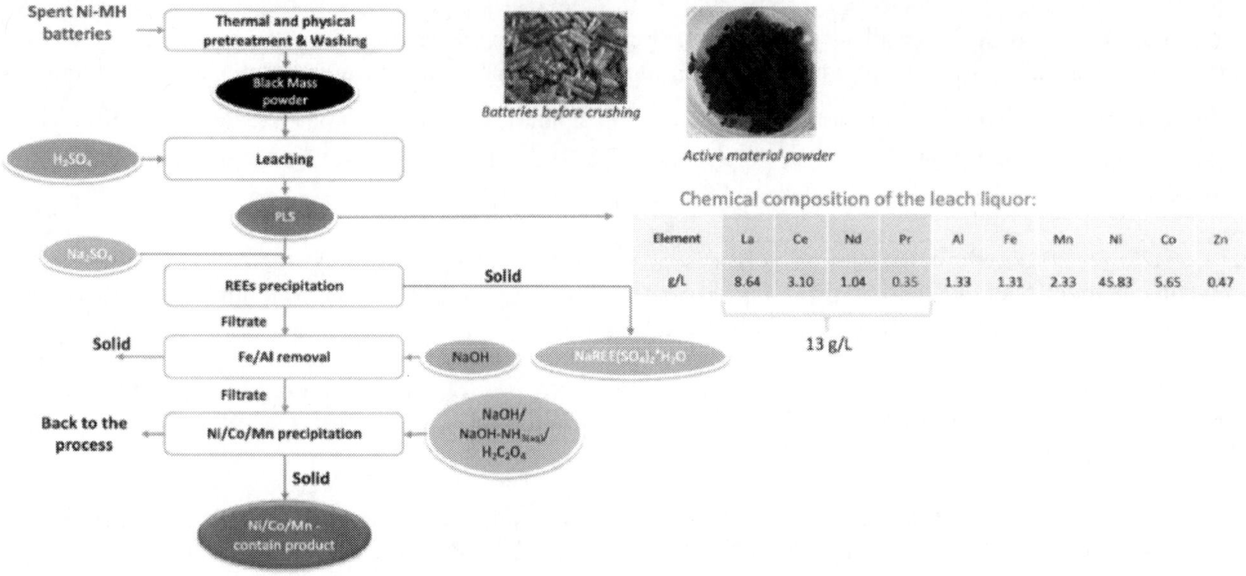

Figure 1: Principle diagram of the treatment of spent batteries powder by successive precipitation processes of valuable metals

	Al	Fe	Co	Mn	Ni	Zn	Sum of REEs
PLS - REEs $g.L^{-1}$	1.30 ± 0.04	1.34 ± 0.05	5.82 ± 0.16	2.35 ± 0.06	45.70 ± 1.77	0.44 ± 0.02	0.59 ± 0.01

Table 1: Chemical composition of the PLS after REEs precipitation

92d Stabilization and Coagulation of Colloidal Suspensions during Crystallization

Xiongtao Ji, Ting Wang, Na Wang, Xin Huang, Hongxun Hao

In colloidal suspension, the disordered colloidal particles could be rearranged under the artificial control, which show great potential in some areas like nanomaterials, energy and biomedicine. However, colloidal suspension could be catastrophic for the phase separation process like crystallization. Due to the high viscosity, colloidal suspension would be a great burden on the equipment, which cause the clogging in pipeline and difficulty in stirring. In addition, due to the collapse of the colloidal suspension, impurities and solvents trapped significantly affect the purity of the final product. Hence, understanding the motion and interaction of particles are crucial and instructive to control the colloid suspension during crystallization. Due to the special thermodynamic properties of cefradine, the mutation of supersaturation usually led to the explosive nucleation, resulting in the formation of large amounts of nanocrystals as the colloidal particles. Via surface analysis of nanocrystal and molecular simulation, the electric double layer on ionized crystal surface was revealed. Then, DLVO model was further derived to analyze interaction between particles. Combining dynamic light scattering, zeta potential measurements and rheology, the stability of colloidal suspension and motion behavior of particles was understood during the whole crystallization process. The developed knowledge based on the suspension can form a basis for further optimization of purification and crystallization for cefradine.

113a Prediction of Nanoparticle Dispersibility Based on Its Surrounding Dielectric Environments

Shota Sampei, Keishi Suga, Taketo Mochizuki, Kanako Watanabe, Daisuke Nagao

Metal oxide nanoparticles exhibit unique optical, electrical, magnetic, and catalytic properties that are different from those of bulk materials, so they have been utilized as fillers to functionalize polymer nanocomposite films. For efficient utilization of such nanoparticles, they must exhibit high dispersibility in media. However, most of the metal oxide nanoparticles possess hydrophilic surfaces, thus they are less dispersible in organic media than aqueous ones with high polarity. The surface modification of nanoparticles is a useful approach to improve the dispersibility of particles, and it should precisely be controlled to disperse the particles in desirable media.

To establish the method for performing the precise surface modification, it is necessary to evaluate the effect of surface modification on the surface properties of nanoparticles, and to quantitatively understand the behaviors of particles in the medium. One approach is to evaluate the Hansen solubility parameter (HSP), which is commonly used to understand the affinity between different substrates (e.g., polymer and solvent). In HSP theory, the cohesive energy density of a substance is divided into three components: dispersion, polar, and hydrogen bonding. Previous studies reported that the HSP values of surface-modified nanoparticles became closer to the ones of surface modifiers compared to unmodified nanoparticles [Ref 1]. However, the quantitative correlation between the amount of surface modifiers bound to the nanoparticles and each component of the HSP value of surface-modified nanoparticles is scarcely clarified. Furthermore, it is necessary to evaluate the dispersibility of nanoparticles in many solvents to calculate the HSP value, which requires lots of time and cost. Therefore, a novel method to evaluate the characteristics of particle surfaces should be established.

In this study, the surface-modified silica nanoparticles (SiO_2) were synthesized using octyltriethoxysilane (OTES) as a modifier. The surface modification ratio, θ, was defined as the ratio of the number of OTES molecules to that of silanol groups on the SiO_2. Then, the HSP values of surface-modified SiO_2 with different θ values were evaluated. In addition, the local polarity in the vicinity of nanoparticles was characterized based on the fluorescent spectroscopy analysis using 6-lauroyl-2-dimethylaminonaphthalene (Laurdan). Laurdan, a fluorescent dye that emits fluorescence depending on the polarity of the local environment, is utilized to evaluate the local polarity at the interface. The surface-modified SiO_2 with higher θ exhibited a high dispersibility in less polar solvents (e.g., *n*-hexane). It was also confirmed that such nanoparticles possessed a decreased surface polarity. Finally, the correlation between the surface properties and the dispersibility was investigated, which reveals that our characterization method can be used to predict the dispersibility of nanoparticles in the medium.

[Ref 1] N. Fujiwara *et al.*, *Mater. Chem. Phys.*, **229**, 139–148 (2019)

113b Magnetic Nanoparticle Characterization and Quantification of Endocytosis Mechanisms By Particle Tracking Velocimetry

Abhinav Sannidhi, Paul W. Todd, Thomas R. Hanley

The properties of individual magnetic particles are of utmost importance in determining their fate in different biomedical and toxicology applications. Utilizing the particle tracking velocimetry in dark-field imaging we were able to quantify the magnetophoretic mobility distributions and intrinsic magnetic properties of individual magnetic particles. In this study, we have quantified the number of superparamagnetic iron oxide nanoparticles (SPIONs) ingested per cell and the kinetics of in vitro receptor-independent endocytosis of different particles with cultured Chinese hamster ovary (CHO-K1) cells. Magnetophoretic mobility of the CHO-K1 cells ingested with different SPIONs is directly proportional to the cellular uptake (beads per cell). The average magnetophoretic mobility for a particular sample is shown to be dependent on the particle surface chemistry, time of incubation, and particle concentration. The dominant endocytic pathway for each particle type has been determined using the specific inhibitors, genistein and chlorpromazine for caveolae and clathrin-dependent endocytic mechanisms respectively. The toxicity of these two specific inhibitors and their effect on CHO-K1 cell viability was quantified by magnetophoretic mobility measurements. This time-efficient magnetic characterization is useful in quantifying the magnetic properties of individual nanoparticles, their cellular uptake, and helps in the selection of particle surface chemistry for cellular endocytosis.

113c Pore Structure Dynamics of Carbonaceous Particles By Surface and Internal Oxidation

Georgios Kelesidis, Patrizia Crepaldi, Sotiris E. Pratsinis

The dynamics and impact of internal and surface oxidation on carbon particle (CP) pore structure are elucidated by lattice Monte Carlo (LMC) simulations and compared to experimental data from the literature (Kelesidis et al., 2022). The fractal-like pore structure or network of unoxidized or as-prepared CPs is determined by diffusion limited aggregation and validated with those measured for carbon blacks and soot. At high temperatures (i.e. 800 °C and above), surface oxidation reduces the CP diameter, d_p, without affecting its internal structure. So, the evolution of CP mass, d_p and specific surface area, *SSA*, closely follows the classic shrinking particle model (Kelesidis & Pratsinis, 2019). At low temperatures (550 °C), internal oxidation dominates, compacting the CP fractal-like pore network. This triples the CP *SSA* up at 75 % conversion. In the presence of a non-reactive graphitic shell though, internal oxidation stops after the reactive and largely amorphous CP core is consumed (Fig. 1) leveling off its *SSA* at large conversions (> 50 %). Accounting for the realistic core-shell pore structure of CPs during internal and surface oxidation nicely explains several measurements of soot and carbon black *SSA*, for the first time to the best of our knowledge. So the LMC-derived oxidation dynamics of CPs presented here can be used to assist the design of highly porous carbon black grades from first principles, as well as the mitigation of soot emissions by enhancing their oxidation in the exhaust of combustion engines.

References:

Kelesidis, G. A., and Pratsinis, S. E. 2019. Estimating the internal and surface oxidation of soot agglomerates. Combust. Flame 209: 493-499, doi.org/10.1016/j.combustflame.2019.08.001.

Kelesidis, G. A., Rossi, N., and Pratsinis, S. E. 2022. Porosity and crystallinity dynamics of carbon black during internal and surface oxidation. Carbon 197:334-340, doi.org/10.1016/j.carbon.2022.06.020.

Figure 1. Cross-sections of CPs derived by LMC for internal and surface oxidation and mass conversions of 0, 25, 50 and 75 %, along with microscopy images of Printex 95 (Kelesidis et al., 2022).

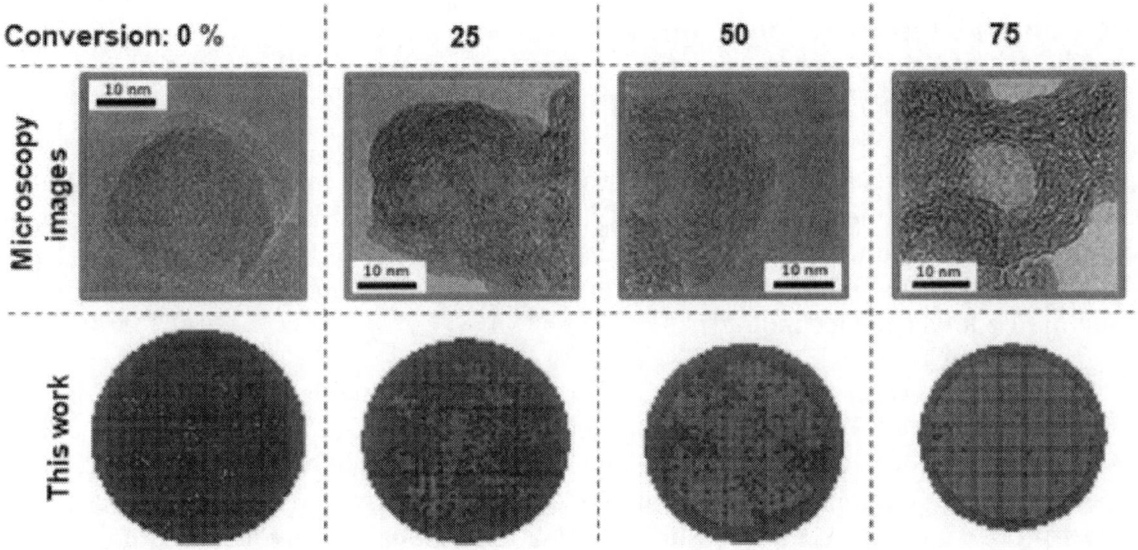
Conversion: 0 %
25
50
75
Microscopy images
10 nm
10 nm
10 nm
10 nm
This work

113d Catalyst Nanoparticle Generation Routes for Deep Injection Floating Catalyst Chemical Vapor Deposition

Jui Junnarkar, Clayton Kacica, Arthur Sloan, Matteo Pasquali, Glen Irvin

Carbon nanotubes (CNTs) are among the most promising candidates to meet the demands of next-generation materials applications, from high-strength light-weight construction to the development of field effect transistors. The future development and adoption of many of these technologies depends on a supply of CNTs that are high-quality (i.e., long and highly crystalline) and affordable enough to enable competition with legacy technologies.

Floating catalyst chemical vapor deposition (FCCVD) is the most scalable method for the production of high quality few walled CNTs due to its continuous nature and lack of a catalyst substrate. FCCVD operates by using an aerosol of catalytic nanoparticles composed of iron or a similar metal suspended in the process gas. These floating particles serve as active sites for CNT growth. This reduces the number of post-processing steps needed to purify CNTs. The FCCVD technique also allows for the direct production not only of CNT powders, but also of fibers, films, and foams directly from the reactor. The recent development of deep injection floating catalyst chemical vapor deposition (DI-FCCVD), in which reactants are injected deep into the deep into the high temperature zone of the reactor has dramatically improved the production rates of high-crystallinity CNTs [1]. Despite these advantages, FCCVD and DI-FCCVD are still limited by poor reactant utilization and low process intensity, leading to high production costs for high-quality CNTs.

Generating a catalyst nanoparticle aerosol flow of the appropriate diameter and flow, with narrow polydispersity, is a critical step in the FCCVD process [2]. Most catalyst nanoparticle production methods rely on the thermal decomposition of organometallic precursors in situ without process control or validation, resulting in poor control over catalyst particle size and composition [3]. Prior observations suggest that poorly controlled particle diameter may contribute to low catalyst utilization and low efficiency in the FCCVD process [3] Catalyst utilization, which we defined as the proportion of catalyst material that successfully grows a CNT, remains below 5 % in current reactors. DI-FCCVD is also limited by the coagulation of catalyst particles. Thus, we hypothesize that the apparent trade-off between high throughput particle synthesis and particle size is due to poor catalyst particle size control.

To date, DI-FCCVD has only been studied with catalyst formed by precursor decomposition. The use of pre-formed catalyst particles offers a potential route to further improve catalyst efficiency in these processes. To this end, we have developed a reliable catalyst generation system using an atmospheric pressure radio frequency plasma in a mixture of argon and hydrogen. Our system decouples the catalyst formation process from the CNT growth process, allowing for process parameters to be individually optimized for both. The high energy density of plasma makes it ideal for the formation of catalyst particles, as it can utilize metallic catalyst sources, such as an iron wire, rather than a precursor molecule, removing extraneous carbon from the reaction. In addition, it is well known that nanoparticles created in plasma are primarily negatively charged which slows particle agglomeration, enabling narrow size distributions [4] We developed a plasma based particle generation system and characterized the effects of process geometry, the effects of plasma power, post-plasma dilution of the particles, and the inclusion of process additives, such as sulfur compounds by scanning mobility particle sizer, transmission electron microscope, and scanning electron microscope. We supplement our empirical characterizations by modeling the plasma in COMSOL. We then coupled this system to a DI-FCCVD process in order to synthesize high quality CNTs. We present our results in optimizing this process to improve productivity, catalyst utilization, and reactant conversion to few walled, high aspect ratio CNTs. We expect this work to serve as a starting point for studies demonstrating the enhancement of control and scalability of catalyst particle delivery for DI-FCCVD reactors.

References Cited

1. Lee, Sung-Hyun, Junbeom Park, Ji Hong Park, Dong-Myeong Lee, Anna Lee, Sook Young Moon, Sei Young Lee, Hyeon Su Jeong, and Seung Min Kim. "Deep-Injection Floating-Catalyst Chemical Vapor Deposition to Continuously Synthesize Carbon Nanotubes with High Aspect Ratio and High Crystallinity." *Carbon* 173 (March 1, 2021): 901–9. https://doi.org/10.1016/j.carbon.2020.11.065.
2. Hoecker, Christian, Fiona Smail, Mark Bajada, Martin Pick, and Adam Boies. "Catalyst Nanoparticle Growth Dynamics and Their Influence on Product Morphology in a CVD Process for Continuous Carbon Nanotube Synthesis." *Carbon* 96, no. Supplement C (January 2016): 116–24. https://doi.org/10.1016/j.carbon.2015.09.050.
3. Zhang, Xiao, Brian Graves, Michael De Volder, Wenming Yang, Tyler Johnson, Bo Wen, Wei Su, Robert Nishida, Sishen Xie, and Adam Boies. "High-Precision Solid Catalysts for Investigation of Carbon Nanotube Synthesis and Structure." *Science Advances* 6, no. 40 (September 30, 2020): eabb6010. https://doi.org/10.1126/sciadv.abb6010.
4. Goree, J. "Charging of Particles in a Plasma." *Plasma Sources Science and Technology* 3, no. 3 (August 1994): 400. https://doi.org/10.1088/0963-0252/3/3/025.

113e Data Driven Method Using Powder Characterisation Tools to Calibrate DEM Simulations

Ben Jenkins, Andrei-Leonard Nicusan, Aurelien Neveu, Geoffroy Lumay, Filip Francqui, Jonathan Seville, Kit Windows-Yule

Background

Powder and particle characterisation tools have been essential to improving our understanding of powders since the 1700s. They have become much more advanced and automated in the 21st century but the way measurements are used remains the same, as a number to compare between powders. To take powder and particle characterisation tools to the next level, the measurements in the lab need to be able to directly inform decisions and solve problems in industrial equipment.

Powder characterisation tools are already advanced enough to be able to make a variety of accurate measurements of bulk powder properties such as the angle of repose. The discrete element method (DEM) has started to allow powders to be modelled on large scales in industrial equipment, improving our ability to design powder handling equipment and solve issues. However, to calibrate DEM simulations the microscopic properties of the powder being simulated are required. A method is needed to calculate the microscopic properties from the bulk measurements made using powder characterisation tools.

Finding the relationship between the microscopic properties and measurable bulk macro measurements, a micro-macro relationship mapping, would allow powder characterisation tools to measure the properties of the particles that make up the powder. This would also provide a better calibration method than currently used methods which can be inaccurate and time intensive to conduct.

Method

Bulk measurements from 4 powder characterisation tools: GranuDrum, GranuHeap, GranuFlow and GranuPack will be linked to the microscopic properties: restitution, sliding friction, rolling friction and cohesive energy density, using a data driven method. To generate the data required, DEM simulations of each powder characterisation tool (digital twins) were developed. Thousands of simulations of each digital twin were then conducted on supercomputers and the bulk measurements recorded at various different values of microscopic properties.

To develop the micro-macro relationship, the typically data-driven approach today would be to train a neural-network. However, this has various problems including poor extrapolation and the fact that neural networks could fit anything to anything. Instead, an exploratory equation approach is being taken. This is similar to fitting a simple second order model to a dataset but instead of fitting just the coefficients you also generate a unique equation structure. This has the advantage of having better extrapolation due to trying to fit a physical equation to the data and reduces the black box nature that neural-networks have as you end up with a real equation you can manipulate afterwards.

Results

Apart from the novel calibration approach developed as part of the project that allows the calculation of microscopic properties of particles from quick powder characterisation tools experiments, digital twins have been developed and sensitivity analysis have been conducted. The top half of the figure below shows the digital twins that have been developed. From left to right is the GranuHeap, GranuPack, GranuFlow and GranuDrum.

Using the data generated for the data-driven calibration method there are other analysis techniques that can be used to provide further insight into powder characterisation. For example, high-dimensional model representation sensitivity analysis was used to determine how influential each microscopic particle property is on the bulk measurements made by powder characterisation tools. The results are shown in the lower half of the figure below.

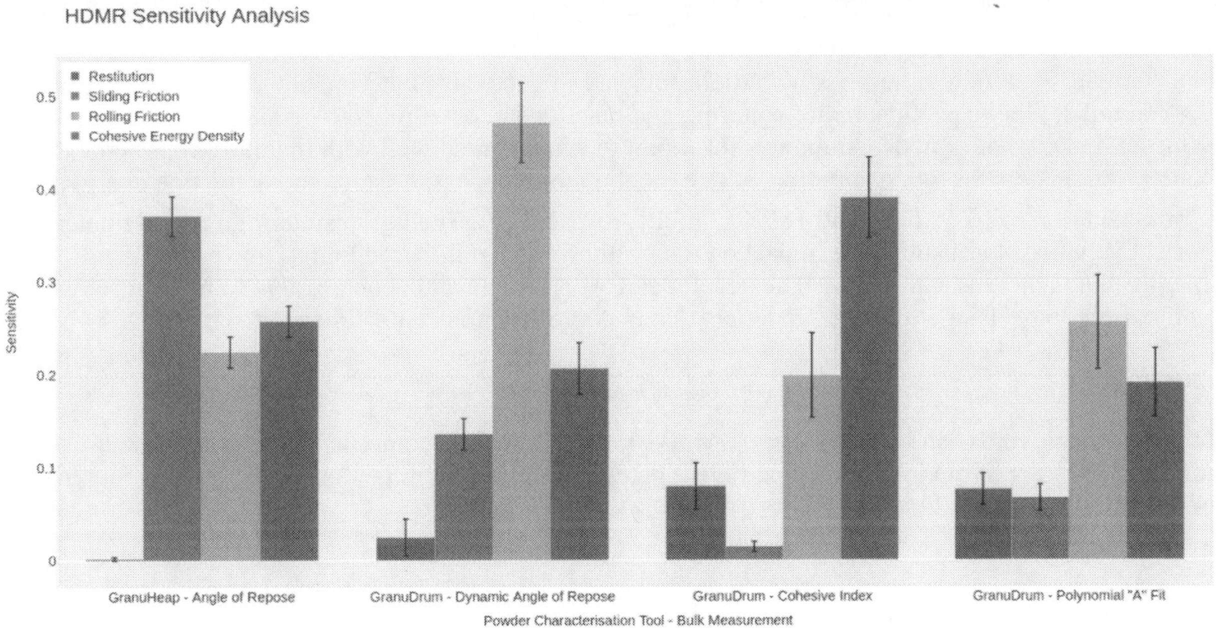

113f Modeling Size and Morphology Evolution of Multiparticle Aggregates during Simultaneous Reaction and Sintering

Sujit Modi, Onochie Okonkwo, Hao Zhou, Shalinee Kavadiya, Marcus Foston, Pratim Biswas

The size and morphology of nanosized aggregates play a crucial role in determining their performance in several applications, including healthcare, energy storage, and catalysis. Thus, knowledge and control of processes that determine the primary particle size and morphology of aggregates is crucial to synthesizing nanomaterials of optimal performance. Both the size and morphology are impacted by several aerosol mechanisms, including sintering. Many systems in which sintering occurs also undergo concomitant inter/intra-particle processes such as coagulation, growth, and chemical reaction. Numerical models have successfully addressed this challenge for scenarios involving simultaneous coagulation and sintering as well as growth and sintering; however, to the best of our knowledge, scenarios with simultaneous reactions and sintering have not been addressed.

In the present study, a novel geometric model (GM) is developed to predict the evolution in size and morphology of multiparticle aggregates under simultaneous sintering and chemical reactions. The GM enables the estimation of sintering kinetics to predict changes in primary particle size, neck size, and surface area of aggregate as a function of residence time and temperature. Lignin, the second most abundant aromatic biopolymer, is used as a model compound and a furnace aerosol reactor system is then used to study the evolution of lignin nanoparticles that are impacted by sintering and reaction. Using the developed model, kinetic parameters for sintering and reaction are determined by comparing GM to the experimental results. The kinetic parameters for lignin reaction agreed well with literature-reported values. The kinetic parameters for lignin sintering, which are the pre-exponential factor and activation energy, were estimated as 6.6×10^{-8} $s.nm^{-1}$ and 116.4 $kJ.mol^{-1}$, respectively. The lignin sintering rate parameters were effectively used to establish the impact on the synthesis of lignin-based high-value products, specifically nanomaterials and bio-oil. The developed GM is simple and generalizable to investigate the size and morphology changes of other materials that undergo reactions with sintering.

Reference:

S. Modi, O. Okonkwo, H. Zhou, S. Kavadiya, M. Foston, P. Biswas, Geometric Model for Predicting the Size and Morphology Evolution of Multiparticle Aggregates during Simultaneous Reaction and Sintering, Chem. Eng. J. (2023) 141423.

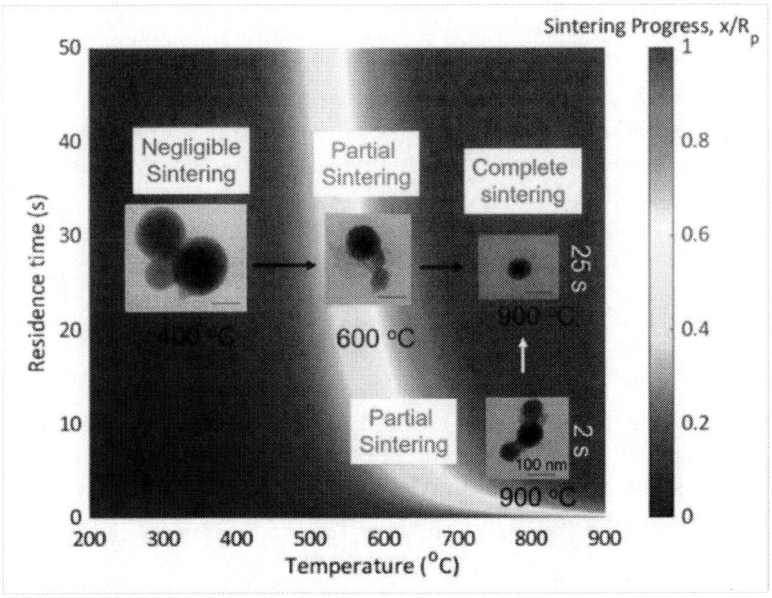

Predicting agglomerate morphology change under simultaneous sintering and reaction

Application: *Controlled synthesis of nanomaterials*

53a Dynamical Analysis of Velocity Time Series in a Binary Fluidized BED: MODEL Validation with Radioactive Particle Tracking (RPT) Experiments

Subi Nath, Akshat Yadav, Shantanu Roy

1. Introduction

Fluidized Bed owing to numerous advantages are extensively utilized in industrial processes. However, due to its complex hydrodynamics and lack of complete understanding, accurate scale up such units is challenging. Therefore, gaining more insights into the complex hydrodynamics of such system is the need of the hour. Euler-Lagrange Model/Discrete Particle Model (DPM), arguably gives more physically realistic results when compared to Euler-Euler model. The present study seeks to capture the complicated flow pattern of such beds using Discrete Element Model (DEM). For validation of the efficacy of the simulations, a direct comparison is made with a experimental technique that is Lagrangian in nature, *viz.* Radioactive Particle Tracking (RPT) (Roy et al., 2021).

Real fluidized beds are polydisperse in nature, and hence it is important that the polydispersity is captured by the DEM simulation technique. While a full treatment of polydispersity is beyond the scope of this work, this work focuses on the study of a binary fluidized bed – examining its hydrodynamics both via CFD-DEM and RPT.

For complete understanding of hydrodynamics of fluidized beds, along with time-averaged data, dynamic analysis of the data is equally important. It is an important tool used in various fields of engineering to study the behavior of systems that evolve over time. Dynamical analysis helps to design, optimize and control complex systems, as well as to understand the underlying physical mechanisms and interactions that govern their behavior. Therefore, study of both time-averaged and time-resolved data is of utmost importance. For time-resolved analysis, Fast Fourier Transform (FFT) is employed which comes under the category of Frequency Domain Analysis. At the later part of the project, state space analysis i.e., Chaos Analysis is also employed to capture the spacio-temporal patterns encountered in a Fluidized Bed.

2. Methodology

2.1 CFD-DEM simulation of a laboratory-scale Binary Fluidized Bed

CFD-DEM simulation of a laboratory-scale 3D binary fluidized bed has been carried out consisting of glass beads of 0.5- and 2-mm diameter particle sizes. To carry out these simulations opensource software package OpenFOAM-CFDEM-LIGGGHTS is used. With the aim of determining the contact forces between the particles, Hertzian model was used. Effect of bed composition was checked on overall hydrodynamics of the system by varying the coarser particle fraction from 10 – 40% at different superficial gas velocities.

Particle governing equations

The governing equation for the translational and rotational motion of a particle can be written as:

(2)

Fi,p , *Fi,w* are the contact force due to particle-particle and particle-wall interactions respectively. *Fi,c* refers to the cohesive forces between particles. The contact forces (normal and tangential) are determined using the following equations. The accuracy of DEM simulation depends upon the parameters: spring stiffness (k_n), damping coefficient (), coefficient of friction (μ). Therefore, selection of these parameters has been done after extensive literature survey.

(4)

(5)

If the following relation is satisfied, then sliding occurs and the tangential contact force is obtained using equation (7),

(6)

(7)

Gas phase governing equations:

Continuity equation

(8)

Momentum equation:

(9)

Where S_p is the source term for the drag force due to the interaction between the gas-phase and particulate phase.

(10)

The δ-function ensures that the reaction force acts as a point force at the position of the particle in the system. V represents the volume of the cell, V_a is the volume of the particle.

- Fast Fourier Transform

Post processing of the simulation data (i.e., instantaneous velocity of the particle) is carried out using Fast Fourier Transform (FFT). It uses a recursive approach that breaks down the signal into smaller sub-signals, computes the Fourier transform of each sub-signal, and then combines the results to obtain the overall Fourier transform of the signal.

Time series of instantaneous velocity measurements from the fluidised bed is collected. It is ensured that the data is properly aligned and normalized to remove any bias or drift that might affect the FFT analysis. The data is also be windowed to reduce spectral leakage. The FFT algorithm is applied to obtain the frequency spectrum of the instantaneous velocity, thus providing information about the frequency content of the signal.

- Chaos Analysis

Chaos analysis is used in the later part of the project to study the behavior of fluidized bed that exhibits unpredictable and sensitive dependence on initial conditions. Parameters like the Kolmogorov entropy and Correlation Dimension are calculated both from the DEM simulations as well as the RPT data.

To estimate the Kolmogorov entropy, two points on the attractor that are closer than a certain length scale is followed in time, until their distance has grown larger than the length scale. By choosing number of such pairs, the entropy is determined with high precision. The Kolmogorov entropy increases continuously with superficial velocity for a fixed aspect ratio of the system. The other parameter calculated is the correlation dimension, which expresses the amount of space occupied by the attractor.

3. **Results**

Figure 1., shows the time-averaged axial mean velocity and time-averaged axial RMS velocity of binary mixture containing 10% 2 mm and 90% 0.5 mm glass bead particles. This simulation was carried out at 1.1 times minimum fluidization velocity employing different drag models Gidaspow (Gidaspow, 1994), Di Felice (Di Felice, 1994) and Koch-Hill (Koch et al., 2001).

The final presentation will discuss these results as well as many other to elucidate the validation of CFD-DEM results with RPT.

4. Conclusions

The observations from simulation were that due to increase in jetsam fraction (2 mm), the mean velocity of flotsam fraction (0.5 mm) was reduced. A reduction in solids fluctuations was also seen both in time-averaged and time-resolved profile. As the bed consists of a distribution of particle sizes, the mixing pattern of the polydisperse bed is studied.

References

Di Felice R. (1994). The voidage function for fluid-particle interaction systems. *Int J Multiph Flow*, 20(1), 153-159.

Gidaspow, D., (1994). Multiphase Flow and Fluidization: Continuum and Kinetic Theory Descriptions. *Academic Press*.

Koch, D.L. and Hill, R.J. (2001). Inertial effects in suspension and porous-media flows. *Annual Review of Fluid Mechanics*, 33, 619-647.

Roy, S., Pant, H.J., & Roy, S. (2021). Velocity characterization of solids in binary fluidized beds. *Chemical Engineering Science*, 246, 116883.

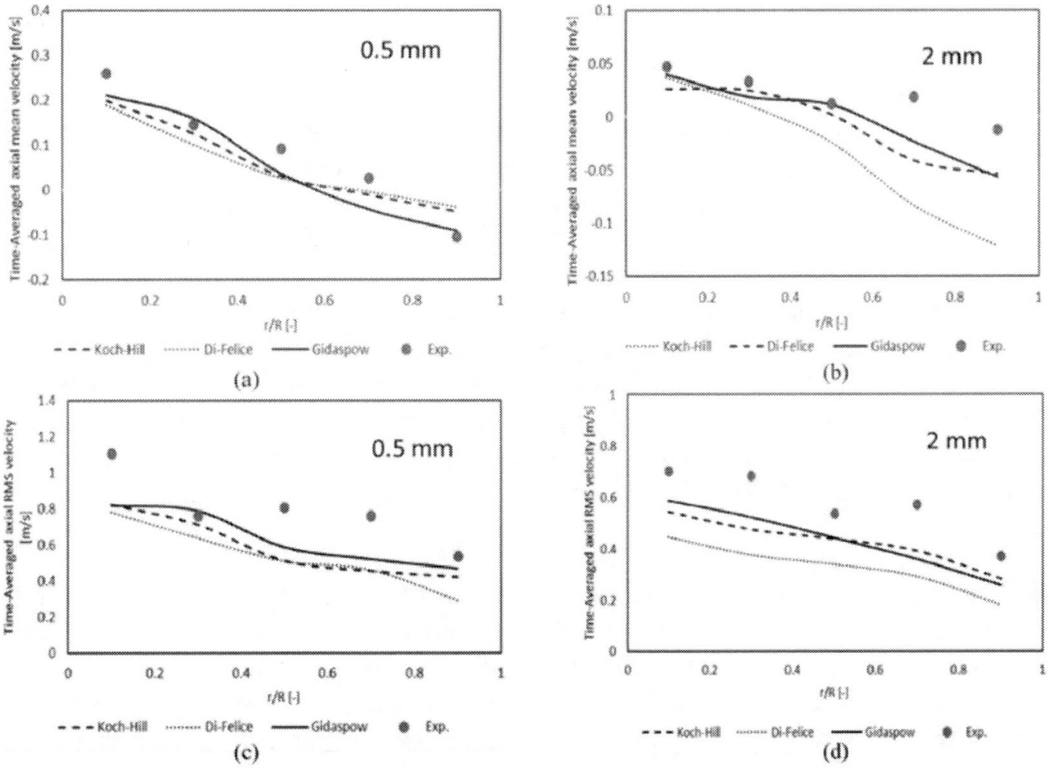

Figure 1. Time-Averaged axial mean Velocity and RMS velocity of 10 % coarser fraction (2 mm) with 1.1 times the minimum fluidization velocity

53b Novel Pathways for Fluidized Bed Coating with the Aid of Non Conventional Fluidizing Media: Experiments and Models

Aydin Sunol, Anurag Guha

Non traditional fluidization media such as supercritical fluids offer novel operational avenues and schemes such as temperature swings for nanolayer deposition while avoiding environmentally questionable solvents. Experimental results from select pharmecautical and energy applications will be introduced and coupled with modeling. A robust model involving dynamic stochastic mass and energy balances for a fluidized bed with or without coating solution streams, where balance equations partitions the fluidized particle bed into well mixed control volumes in the axial direction, treating the fluidization fluid as plug flow and the particle bed as a CSTR is developed. There are two types of control volumes dependent on the penetration depth of the atomized coating fluid (if any), coating control volumes and non-coating control volumes – each with its own set of differential equations. The results from the dynamic mass and energy balances include: temperatures for (fluidization fluid, particles, and fluidized bed wall), solvent content (fluidization media and particle), and the coating mass balance.

The coating growth kinetics model developed for this work is based on the fluidized bed hydrodynamic properties (bubble diameter, velocity, and void fraction, fluidized bed void fraction, and particle circulation rate) calculations from the aforementioned results. The hydrodynamic properties are then extended to establish probabilities for being in the spray area and being coated in the spray area. In the absence of data, particle size distributions and droplet size distributions are generated from appropriate relationships (e.g. triangular distribution, Rosin-Rammler distribution). Mass balances for the droplet collisions with the particles in the spray area are tabulated to determine how much coating material is applied during each successful droplet-particle collision. The particle size distribution and coating thickness distribution are then produced as a function of time.

The model is used for a portfolio of applications and fluidization using supercritical carbon dioxide with or without additives in different fluidized beds such as wurster or circulating. Global sensitivity analysis is used for identification key parameters and their interaction. Coupled with pilot plant several encapsulated products were modeled to optimize encapsulant characteristics and development of operational policies.

53c Quantifying Powder Cohesivity through Fluidisation Tests

Amalia Thomas, Nathalie Vriend, Alexander Routh

Geldart provided useful classification criteria of powder fluidization behaviour, and a simple yet often effective metrics for its prediction. However, more factors than only the particle density and diameter affect fluidization potential, such as particle surface properties and shape, particularly near the transitions between Geldart groups. In this work we discuss an experimental rheological test for the characterisation of powder fluidization behaviour and for the determination of relevant parameters such as minimum fluidization speed and fluidization quality.

In the first part of this work we aim to understand the mechanisms affecting particle interactions during the fluidization process. Features of typical pressure drop and voidage curves formed during the fluidization and de-fluidization of powders are compared for a range of powders with different particle sizes and flow behaviours. A hysteresis in the pressure drop across the powder bed, an overshoot in pressure drop around the minimum fluidization air speed, and the fraction of the bed weight suspended by drag are correlated to the observed fluidization behaviour. We discuss the interpretation of such features in the resulting pressure-drop and bed height curves from Fluidization tests carried out in an FT4 Powder Rheometer, where a standard Aeration Test was modified to track the pressure drop and the bed expansion at appropriate air speed intervals.

The ratio of cohesive forces to average particle weight is proposed as a main indicator for the type of interparticle interactions in the powder during fluidization. Based on the hysteresis in pressure drop and voidage curves, we propose a novel method for experimentally determining the Bond number, i.e. the ratio of average interparticle force magnitude to particle weight. The difference between the fluidization and de-fluidization curves is related to the force required to break interparticle forces in the powder bed between two packing states. By computing the change in coordination number (mean number of contact points per particle) estimated from the bed void fraction, the number of interparticle forces broken during aeration is related to the average particle-particle force.

We show results of Fluidization tests for powders of a range of particle sizes and fluidization behaviours. The results are compared to Geldart's classification, to theoretical expectations and also to alternative rheological measurements commonly used in industry, including the Hausner ratio, the angle of repose shear cell cohesion and flow function values, and FT4 Energy measurements. Furthermore, the proposed method is shown to be independent of bed dimensions and initial packing state.

53d 100 Years of Scaling up Fluidized Bed and Circulating Fluidized Bed Reactors

Jia Wei Chew, Wyatt Casey LaMarche, Ray Cocco

The year 2022 marked the 100th anniversary of the commercialization of the fluidized bed reactor. In those years, many new processes have been developed, with many considered breakthrough technologies, replacing technologies that were no longer considered competitive. Fluidized beds have the advantages of superior heat transfer and the ability to continuously move solids during operation, along with a list of other attributes. As a result, processes spanning coal and biomass gasification, pyrolysis, fluidized catalyst cracking, acrylonitrile, polyethylene, oxychlorination, and polycrystalline silicon have prospered with the application of fluidized bed technology. Expect that list to continue with current efforts in chemical looping, plastic pyrolysis, methane pyrolysis, and propane dehydrogenation, to name a few.

During that same period, we saw engineering go from an incremental approach to a more fundamental one using sophisticated models and relevant cold-flow experimentation. That engineering work process is still changing. The onset of better computational platforms, models, and experimental techniques, combined with better statistical tools (e.g., machine learning), leads to better understanding of the underlying physics and enhanced control systems (e.g., artificial intelligence). This will allow for the application of fluidized bed technology using less capital and less operating costs, and also enable commercialization in less time.

53e Behaviour of Lumps in Gas-Solid Fluidized-Bed Reactors

Matteo Errigo, Massimiliano Materazzi, Paola Lettieri

Gas-solid fluidized-reactors have been historically implemented in a variety of fields, with applications to the energy, petrochemical and pharmaceutical, to mention just few. More recently, fluidized-bed reactors have also been employed in waste-to-energy conversion plants with CO_2 capture. The widespread use of such reactors is mostly due to their excellent mixing properties, allowing, among other things, for temperature uniformity and for excellent contact between the gas and solid phases. These properties, however, are not guaranteed if lumps or agglomerates are present the fluid bed. This is the case of lumps, that can form, for example, in industrial applications of thermochemical conversion of waste plastic, as this latter tends to melt and embed bed particles. When the formation of these lumps cannot be avoided, a thorough understanding of their behavior is fundamental to limit the disruption they may cause to the reactor's operation. The aim of this study is to assess the impact of design parameters and operating conditions on the motion characteristics and thermal behavior of such lumps. This knowledge can be helpful in finding the optimal conditions for the prevention of hotspot formation and defluidization.

The experimental setup used in this study consisted of a downscaled pseudo-2D fluidized bed that enabled the direct observation of the simulated lumps via infrared thermography. The fluid bed was operated at ambient temperature and the fluidization velocity was varied between $1U_{mf}$ and $10U_{mf}$, ranging from incipient fluidization through bubbling and slugging. The bed material used was rutile sand; 4 different particle mean diameters, 60μm, 100μm, 153μm and 215μm, allowed to assess the effect of the particle size on the motion and heat transfer characteristics of agglomerates. Artificial lumps were produced by inserting heavier lead tracers into pumice beads; changing the tracer's mass allowed to vary the density of these lumps and better simulate their behaviour in a fluid-bed of sand used for waste plastic treatment. The lumps were heated up and introduced in the fluidized bed. They were then observed as they moved around and exchanged heat with the fluidized bed. X-ray particle tracking (XPT) was used to track the lump position in the fluidized bed. Post-processing of the x-ray images allowed to identify the lump segregation behaviour for different conditions and to obtain the vertical and horizontal dispersion coefficients of the lumps. Infrared thermography (IRT) was used to determine the lump temperature evolution in time. This was done by converting the infrared signal emitted by the simulated lump into a temperature via a calibration curve that had been previously obtained. The heat transfer coefficient between lump and fluidized bed was then obtained through an energy balance.

The effect of the lump density on the lump dispersion coefficients was found to be negligible. Larger fluidization numbers and larger particle sizes, on the other hand, were both found to increase the dispersion coefficients of the lumps following power laws, as shown in Figure 1a,b. The lump segregation behaviour was then investigated. The lump density only plays a role for small fluidization velocities, up to $2U_{mf}$ for the bed material with particle size equal to 215 μm. Depending on its density, in fact, the lump was seen to sink at the bottom of the fluid bed, floats at the top or is properly mixed within it, as shown in Table 1a. It was concluded that beyond a threshold value for the fluidization velocity, $2U_{mf}$ in the case studied, mixing overcomes segregation, which is due to the difference in size and density between the lump and the bed material. Furthermore, Table 1b shows that a larger particle size allows better lump mixing at smaller fluidization numbers. Once more, the effect of the density of the lump on its heat transfer coefficient with the fluidized bed is negligible, as shown in Figure 2a. On the other hand, an increase in the fluidization velocity caused an increase in the heat transfer coefficient (Figure 2a,b), but only up to a point. The heat transfer coefficient, in fact, then stabilizes onto a constant value as the fluidization velocity keeps increasing. This behaviour agrees with previous results in the literature; this was attributed to the fact that larger gas flow rates improve the mixing and decrease the refreshment rate of bed material around a lump. This mechanism improves the heat transfer but soon saturates, hence the heat transfer follows an asymptotic trend with respect to the fluidization velocity. On the other hand, the heat transfer coefficient trend as a function of the fluidization number strongly depended on the bed material particle size, with the maximum value for each particle size decreasing with the bed material particle size, as shown in Figure 2b and in agreement with previous research. Due to the large size difference between the bed material and the lumps, particle convection was expected to be the main heat transfer mechanism. Once the bed starts bubbling, as the gas flow rate through the fluid bed increases, the excess gas ends in bubbles and the gas superficial velocity through the emulsion phase remains approximately constant. However, the heat transfer coefficient was observed to sharply increase in this range. This was interpreted as proof that particle convection is the dominant heat transfer mechanism. Group A particles presented a first very small increase in the heat transfer coefficient for superficial velocities between the minimum fluidization and the onset of bubbling. In this range of velocities, the whole gas flow rate ends up in the emulsion phase, but this has a negligible effect on the heat transfer coefficient. This is a further proof that particle convection is the main heat transfer mechanism. In fact, as soon as the bed starts bubbling and the mixing within the bed improves, the heat transfer coefficient increases sharply by a factor up to 10 (for average particle diameter equal to 60 μm). Optimal values of fluidization velocity that would guarantee sufficient mixing to prevent segregation, while also maximizing the heat transfer coefficient were found for different configurations. Going beyond these values returns negligible advantages from the points of view of lump mixing and heat transfer. Instead, it may increase the chance of operational problems, such as attrition and elutriation.

While semiempirical correlations for the Nusselt number between larger objects and fluidized bed have already been obtained for gas-convection dominated scenarios, this is not true for cases where particle convection is the main heat transfer mechanism. Hence, a correlation for the Nusselt number as a function of the Reynolds number and of the lump-to-bed-material size ratio was derived (Figure 3):

$$Nu^* = 2 + 0.037\, Re_{lump}^{0.38} \, (d_{lump}/d_p)^{1.14}$$

where the object Reynolds number, Re_{obj}, was defined with the lump velocity information obtained through XPT.

X-ray particle tracking and infrared thermography were used to characterize the motion characteristics and thermal behaviour of lumps in a fluidized bed. Such lumps can impact the quality of fluidization and cause the formation of hotspots. A thorough knowledge of their behaviour is hence very important for industrial applications of fluidized-bed reactors. It provides, in fact, an insight into how process variables may affect the behaviour of such lumps, helping in preventing the disruption of such reactors.

Table 1. Segregation behaviour of artificial lumps (a) for different values of fluidization number and agglomerate-to-particle density ratio ($d_{p,mean}$=215μm) and (b) for different values of fluidization numbers and bed material particle size (ρ_{obj}/ρ_p=0.32).

(a)

	U_{mf}	$1.5U_{mf}$	$2U_{mf}$	$4U_{mf}$	$6U_{mf}$	$8U_{mf}$	$10U_{mf}$
ρ_{lump}/ρ_p=0.32	Float	Float	Float	Mixed	Mixed	Mixed	Mixed
ρ_{lump}/ρ_p=0.40	Float	Float	Mixed	Mixed	Mixed	Mixed	Mixed
ρ_{lump}/ρ_p=0.47	Float	Float	Mixed	Mixed	Mixed	Mixed	Mixed
ρ_{lump}/ρ_p=0.55	Sink	Sink	Sink	Mixed	Mixed	Mixed	Mixed

(b)

	U_{mf}	$1.5U_{mf}$	$2U_{mf}$	$4U_{mf}$	$6U_{mf}$	$8U_{mf}$	$10U_{mf}$
$d_{p,mean}$=60μm	Float	Float	Float	Float	Float	Mixed	Mixed
$d_{p,mean}$=100μm	Float	Float	Float	Float	Mixed	Mixed	Mixed
$d_{p,mean}$=153μm	Float	Float	Float	Mixed	Mixed	Mixed	Mixed
$d_{p,mean}$=215μm	Float	Float	Float	Mixed	Mixed	Mixed	Mixed

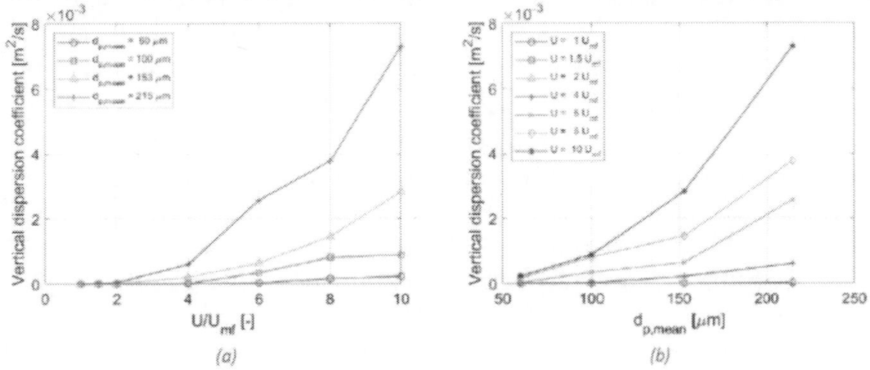

Figure 1. Vertical coefficients of an artificial lump (a) versus fluidization number for different bed material particle sizes (ρ_{obj}/ρ_p=0.32) and (b) versus bed material particle size for different fluidization numbers (ρ_{obj}/ρ_p=0.32).

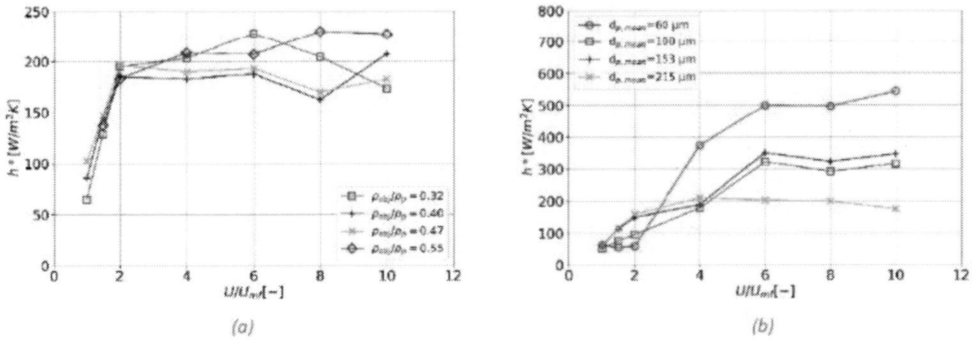

Figure 2. Heat transfer coefficient between bed and artificial lump (a) versus gas superficial velocity for different values of lump density ($d_{p,mean}$=215μm) and (b) versus superficial gas velocity for different values of bed material mean diameter (ρ_{obj}/ρ_p=0.32).

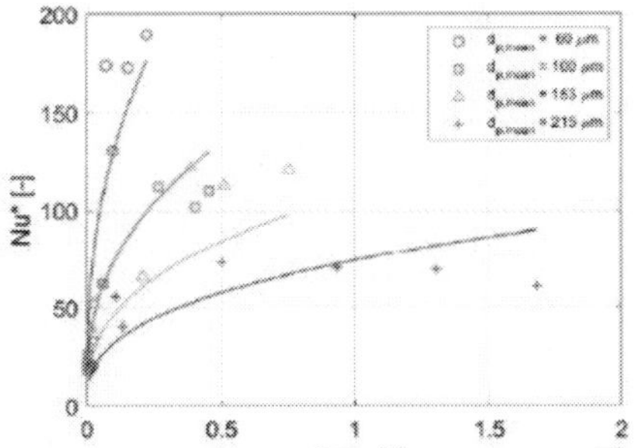

Figure 3. Apparent Nusselt number plotted versus object Reynolds number for different bed material particle sizes: experimental data (points) and fitting (solid lines).

53f Improving FCC Regenerator Performance and Capacity Utilization through CFD Analysis

Zhizhong Ding, Todd Foshee, Manoj Katakdaunde, Robert Ludolph

Fluidized Catalytic Cracking (FCC) plays a significant role in crude oil refining, converting large oil molecules into economically favorable smaller molecules that will satisfy the daily commuting and transportation needs of consumers. Full utilization of the FCC capacity is critical for meeting the daily production requirements from the refinery. The FCC regenerator can be the source of capacity underutilization if it is negatively impacted by the performance of its internals. Internals that promote uniform catalyst distribution are crucial for achieving efficient, reliable spent catalyst regeneration and thus consistent quality catalyst circulated to the reaction system. Regeneration constraints are in turn mitigated (like afterburning) and the lifespan of equipment prolonged.

Shell utilizes Computational Fluid Dynamics (CFD) technology to assess and improve conventional and transitional FCC operations. CFD has become an indispensable engineering tool to optimize the equipment design as well as study the fluidization behavior over a wide range of applications. To illustrate the value delivered by CFD, we will review the performance and mechanical integrity of an existing spent catalyst distributor in an FCC regenerator along with the CFD model created to benchmark its performance. A second CFD model that incorporates a Shell-designed spent catalyst distributor will also be presented and compared to the existing design, focusing mixing, catalyst distribution, and flow behavior. The upgraded distributor was shown to be advantaged in catalyst distribution with improved mechanical integrity, this providing the refiner the opportunity to increase FCC utilization and optimize unit operation overall.

103a Interaction of Granular Bubbles and Droplets in Gas-Bubble-Free Fluidized Beds

Qiang Guo, Christopher Boyce

Recent studies have found that assemblies of granular particles lighter than others in a gas-bubble-free fluidized bed tend to rise as a unit or "bubble", while assemblies of denser particles or "droplets" tend to undergo binary splitting events as they descend. Here, we use computational fluid dynamics – discrete element method (CFD-DEM) simulations to investigate the dynamics of granular bubbles and droplets in which the particles are strongly adhered together, so that they rise or descend as a rigid body. We study the effects of particle size and density ratios on dynamics, as well as the interaction of multiple bubbles or droplets rising or descending together.

103b Fluidization Regimes in High Flux-High Velocity CFB Risers

Allan Issangya, S. B. Reddy Karri, T. M. Knowlton, Jia Wei Chew

Circulating fluidized beds (CFBs) have applications in gas-solids reaction and gas-phase catalytic reaction processes, predominantly in fluidized bed coal combustion and fluid catalytic cracking (FCC) of crude oil and to a lesser extent in calcination. CFBs use is also expanding to other areas such as dehydrogenation and chemical looping. There is, therefore, still need for continued study of CFB riser hydrodynamics, especially as it applies to fast reactions processes. That calls for studies at high superficial gas velocities and solids fluxes, and to be more relevant to industry, conducted in large-scale CFB systems. Additionally, computational fluid dynamics (CFD) modelling is increasingly being adopted by industry. Experimental data are needed for validation of the CFD models. Tests were conducted in this study with FCC catalyst particles in a 0.3-m diameter CFB riser operating at a superficial gas velocity of 12 m/s and solids fluxes of about 300 to 700 kg/sm^2. Fluidization behavior in the riser, assessed from axial voidage profiles, local solids fluxes and high-speed video imaging, suggested a co-existence of the dense suspension upflow regime (Grace, 2000) and dilute pneumatic transport regime. There was no solids refluxing at the wall.

References:

John R Grace (2000), Reflections on turbulent fluidization and dense suspension upflow, Powder Technology, 113 (3), 242-248.

103c Numerical Study on Clustering of Polydisperse Group-A Particles in Riser Flow

Sathvik Bhat, Pedram Pakseresht, Yuan Yao, Yi Fan, Jorg Theuerkauf, Jesse Capecelatro

Particle clustering plays a crucial role in the performance of circulating fluidized bed (CFB) risers. Over the past few decades, numerous studies have been conducted to characterize clusters and their effect on mixing, with the majority of them focusing on monodisperse particles. In this work, numerical simulations of riser flow made up of a continuous distribution of Geldart Group-A particles are conducted. Particle diameters are sampled from a lognormal size distribution, informed from dynamic image analysis of zeolite particles, spanning 21 to 140 um with a mean of 58 um. Highly-resolved simulations of fully-developed particle-laden pipe flow are conducted using the open-source multiphase flow solver NGA2. The volume-filtered Eulerian-Lagrangian equations are coupled with an immersed boundary method to efficiently handle the gas-particle flow. Models are employed to capture subgrid-scale velocity fluctuations induced by particles (termed pseudo-turbulence). The flow begins perfectly mixed and relatively uniform. Strong momentum coupling between the phases results in the spontaneous generation of dense clusters that fall near the riser walls (annulus). It is hypothesized that smaller particles are more likely to rise fast with the high-speed flow in the relatively dilute core, while larger particles are more susceptible to gravity and remain in the slower moving dense annulus. Size segregation during the initial transient will be quantified. Two-phase flow statistics at steady state will be studied to reveal correlations between voidage and velocity with particle size. Finally, a conclusion is drawn on the effect of polydispersity on the size and the shape of the clusters.

103d Scaling with Microstructure Dynamics

Ronald Breault

Scaling circulating fluidized bed risers has been a point of contention for nearly a century. There have been numerous attempts to define various methodologies. These have all fallen short of providing a robust approach, primarily due to a lack of maintaining microstructure dynamics – a key factor in maintain interphase heat and mass transfer. Recent work by the author put forward an approach based up maintaining dynamic similarity at different scales by ensuring that the microstructure remained similar by maintaining statistical and chaotic parameters across the scale. That work relied on a dimensionless regime map in which the x-axis was defined by the ratio of the solids flux to the saturation carrying capacity. This latter property was only good within the range of the data and it became evident that extrapolation beyond the limits of the data could induce unrealistic conditions.

Therefore, literature was reviewed to develop a better correlation for the saturation carrying capacity. In doing so, a critical diameter concept was developed and applied to the correlations for both Geldart Group A and B materials. The critical diameter for Geldart Group A and B materials is 0.2 m and 0.3 m, respectively. The paper then gives three examples on how to use the scaling approach to maintain dynamic similarity across the scales.

103e Validation of New Scale-up Methodology for Gas-Solid Fluidized Bed Reactor Using Advanced Measurement Techniques*

Abdelsalam Efhaima, Al-Dahhan Muthanna

This study focuses on validating the new mechanistic scale-up methodology for hydrodynamics similarity of gas-solid fluidized bed reactors that has been developed in our laboratory. It is based on maintaining similar or closer time averaged radial profiles of gas holdups in two different gas-solid fluidized beds to achieve local and global similarity of the dimensionless hydrodynamic parameters. This is because the gas dynamic dictates the hydrodynamics of these beds. The new scale-up methodology has been successfully validated by assessing for the first time the local hydrodynamic parameters, such as time averaged cross-sectional distributions and radial profiles of gas and solids holdups at different axial levels measured by gamma ray computed tomography (CT) technique and particles velocity field and turbulent parameters (shear stress, normal stresses, and turbulent kinetic energy) measured by radioactive particle tracking (RPT) technique. The experimental results showed that achieving similarity in the radial profiles of the gas holdup in the geometrically similar beds is essential for ensuring closer or similar local and global hydrodynamics similarity. As the differences increase in terms of magnitude and trend in the gas holdup radial profiles between two beds, the differences increase in the detailed hydrodynamics.

210a MFIX-Exa: CFD-DEM Simulations of Thermodynamics and Chemical Reactions in Multiphase Flows

Roberto Porcu, Jordan Musser, Ann Almgren, John Bell, William Fullmer, Deepak Rangarajan

Carbon capture and storage technologies based on chemical looping reactors (CLRs) offer a promising approach to decreasing CO_2 emissions from energy production and various industrial processes. Large-scale commercial deployment of carbon capture technologies requires scaling up from laboratory to industrial systems. This large separation of scales is known to be fraught with issues, particularly for the multiphase reactors at the core of these systems. Computational tools can help bridge this gap; therefore, we created MFIX-Exa [1, 2], a modern, massively parallel code for the numerical solution of chemically reacting gas-solids flows in complex reactor geometries. The fluid is modeled using a low Mach number formulation with a multicomponent ideal gas equation of state which is imposed as a constraint of the velocity field. The fluid equations are discretized using an embedded boundary-aware Godunov scheme with an approximate projection. Particles can be represented using a discrete element method (DEM) or with a lower-fidelity particle-in-cell (PIC) approach. DEM employs a soft-sphere spring-dashpot model and is evolved using a forward Euler method with subcycling. Conversely, the PIC model uses a continuum stress to approximate collision forces on parcels that represent statistical collections of real particles. The fluid and particle models are coupled through a volume fraction field in addition to interphase mass, momentum, and energy transfer. Additionally, MFIX-Exa contains a novel "*PIC to DEM*" application that converts PIC simulation checkpoint files into high-quality initial conditions for computational fluid dynamics-discrete element method (CFD-DEM) simulations. This talk describes the MFIX-Exa model and numerical approach and showcases several verification and validation tests in addition to weak scaling performance results.

[1] Musser, J., Almgren, A.S., Fullmer, W.D., Antepara, O., Bell, J.B., Blaschke, J., Gott, K., Myers, A., Porcu, R., Rangarajan, D., Rosso, M., Zhang, W., Syamlal, M., 2021. Mfix-exa: A path toward exascale CFD-DEM simulations. The International Journal of High Performance Computing Applications. doi: 10.1177/10943420211009293

[2] Porcu, R., Musser, J., Almgren, A., Bell, J., Fullmer, W., Rangarajan, D., 2023. MFIX-Exa: CFD-DEM simulations of thermodynamics and chemical reactions in multiphase flows. Chemical Engineering Science. doi: 10.1016/j.ces.2023.118614

210b Heat Transfer in Dynamically Structured Fluidized Beds

Shuxian Jiang, Kaiqiao Wu, Marc-Olivier Coppens

Inspired by the origin of ordered structures generated by periodically oscillating fluid flows in nature, such as the sandy ripples on beaches and desert floors, it has been proposed to apply an oscillating gas flow at a state around the minimum fluidization velocity to create a "dynamically structured fluidized bed", which manipulates the gas-solid drag flow periodically in time[1]. The pulsed flow introduces order, and the system reaches a specific fluidization state between fixed and fluidized bed. In quasi-2D flat or 3D annular fluidized beds, bubbles rise and form a hexagonal array with bubbles spaced at certain distances, which is different from the chaotic bubble flows in conventional fluidization. This special fluidization state provides opportunities for process intensification. Previous research has investigated the hydrodynamics in dynamically structured fluidized beds and identified the operation of a structured bubbling flow with bubble size and pattern wavelength that could be tuned by changing the oscillation frequency[2]. However, the influences of dynamical structuring on transfer coefficients are still poorly understood.

An essential step towards utilizing dynamically structured fluidized beds for, e.g., heterogeneously catalysed reactions or drying, is the development of a heat transfer model that considers the unusual characteristics of dynamically structured fluidized bed, lying in between traditional fluidized and fixed beds. In this work, the heat transfer characteristics within a 3D annular pulsed fluidized bed were first studied numerically, and subsequently compared with the unstructured conventional bubbling fluidized bed. To evaluate and determine heat transfer performance in this special pulsed fluidized bed system, initially, a systematic analysis was conducted to understand the impact of an oscillating gas flow on three heat transfer mechanisms: particle-particle conduction, particle-fluid-particle conduction, and particle-fluid convection. Then, a computational fluid dynamics discrete element method (CFD-DEM) based model coupled with the modified heat transfer model was developed and applied to simulate heat transfer in dynamically structured fluidized beds. To validate the model, the local heat transfer characteristics in a 3D annular bubbling fluidized bed were investigated experimentally as well. By using thermocouples, the instantaneous temperatures were measured at various locations. It was indicated that local heat transfer coefficients were influenced by the pulsed flow properties. Furthermore, the relationship between bubble patterns and temperature distribution was explored. Such heat transfer models considering the spatiotemporally alternating fluid-like and solid-like conditions are essential for the design and scale-up of dynamically structured bubbling fluidised beds for particle processing and catalytic reactions.[3]

References

1. Francia, V.; Wu, K.; Coppens, M.-O. Dynamically structured fluidization: oscillating the gas flow and other opportunities to intensify gas-solid fluidized bed operation. *Chem. Eng. Process.* **2021**, 159, 108143.

2. Francia, V.; Wu, K.; Coppens, M.-O. On the role of energy dissipation in a dynamically structured fluidized bed. *Chem. Eng. Sci.* **2022**, *248*, 117189.

3. Vandewalle, L.A.; Francia, V.; Van Geem, K.M.; Marin, G.B.; Coppens, M.-O. Solids lateral mixing and compartmentalization in dynamically structured gas-solid fluidized beds. *Chem. Eng. J.* **2022**, *430*(4), 133063.

210c Towards Accurate Modeling of Vibration in CFD-DEM Simulations of Vibrated Gas-Fluidized Beds

Qiang Guo, Jieyu Tian, Christopher Boyce

Recent studies have demonstrated that vibration of gas-fluidized beds can produce structured flow dynamics in granular particles, such as structured bubbling, surface waves and convection cells. Computer fluid dynamics (CFD) simulation of these flows has been key to identifying mechanisms underlying these structured flow phenomena. Simulation of vibration of granular particles using the discrete element method (DEM) can be conducted via vibrating the bottom plate. Incorporating this technique with CFD for modeling gas flow in a vibrated gas-fluidized bed is non-trivial because vibrating the entire CFD grid would be computationally expensive. Existing CFD-DEM simulation works model vibration by either oscillating the gravity or moving a bottom plate made by glued particles. Here, we explore multiple methods for modeling vibrated gas-fluidized beds using CFD-DEM simulations and quantify their accuracy based on direct comparison of simulation of structured granular flows. Importantly, we proposed methods to model vibration by moving a gas-penetrable stl geometry in the bottom and imposing a tangential velocity on the sidewalls. Such methods can model the vibration of the whole bed and are the most accurate for the simulation of vibrated fluidized beds so far.

210d Taylor Vortex Flow in Dry Granular Media

Jagan Mohan Sanghishetty, Lina Baroudi, Christopher Boyce

Classical instabilities such as Rayleigh–Bernard rolls, or Taylor vortices are quite familiar in conventional fluids when subjected to external conditions. But the question is, do these instabilities exist in the granular flows? If they exist, are the underlying causes the same as those in conventional fluids? The shearing of fluids in the cylindrical Couette device is a classic viscometric flow as the stresses are readily related to the rheology of the fluid. These flows are characterized by the non–dimensional Taylor number and have proven critical to evaluating and establishing fluid rheological measurement protocols over years. As we celebrate the centennial of GI Taylor's groundbreaking paper on Taylor vortex flow from 1923, we demonstrate that these instabilities can exist in the dense dry granular flows using Discrete Element Method (DEM) simulations. In the figure below, panel (a) shows r and z–velocity vectors on a theta–averaged section plane and panel (b) shows radially averaged theta–velocity contours on the stretched outer cylindrical surface. We see that there are close similarities and differences between granular and conventional fluid vortices, qualitatively. Quantitatively, we characterize aspect ratio, vorticity, driving torque, and granular viscosity for various system heights filled with glass beads. The role of the external force in the system is found to be crucial. We are motivated by the fact that the stress and kinematics relationship in granular materials is complex. Investigating these granular viscometric flows will aid in building better rheological models.

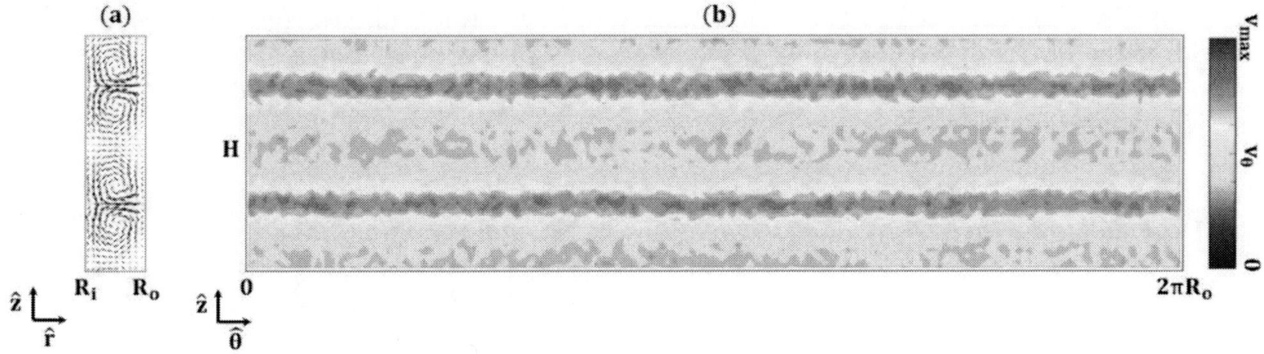

210e Numerical Study of Fluidization of Binary Mixture of Particles Using SOMKTGF-MP

Dan Sun

The second-order moment of kinetic theory of granular flow (SOM-KTGF) has been verified in predicting the particle-laden flow in the median concentration, by the four-way coupled second-order method (FW-SOM) established via coupling the gas phase second-order turbulence model [1]. The anisotropic SOM-KTGF remedied the shortage of the classic isotropic KTGF in predicting the "dilute" phase of particles in the fluidized risers. These particles of the dilute phase in the fluidized bed were suspended in the reactor and less affected by the other particles surrounded. As a result, these particles show high inertia and high anisotropic behavior in the fluctuation of the particle velocity. In the previous study, the FW-SOM and SOM-KTGF showed superior to the classic KTGF in the numerical study of fluidization.

A novel model of the SOM-KTGF for multi-type particles (SOM-KTGF-MP) was proposed in this study to investigate the fluidization of the particle mixture with binary types. The particles in the mixture differ in all types of physical properties such as density, size restitution coefficient et al. In the model, SOM-KTGF-MP, the conservation equations of the volume fraction, velocity, and the velocity fluctuation of particles were solved for each type of particle. In this study, SOM-KTGF-MP was applied to simulate the flow behavior of gas-solid flow with different sizes and densities of particles in the fluidized bed. The anisotropic fluctuation of the binary types of particles and the effects of the particle-particle interaction is investigated in terms of the particle distribution and turbulent characteristics.

[1] D. Sun. Four-way coupled second-order method of two-fluid model for gas-particle flow and the numerical simulation in horizontal channels. Powder Technology 390 (2021) 354–368

210f A General Approach for Modeling Fluidized Beds and Slurries Using CFD-DEM Simulations

John Thomas, Brian DeVincentis

Modeling the behavior of fluidized beds and slurries is a critical aspect of chemical and process engineering, with applications ranging from pharmaceutical production to mineral processing. Computational fluid dynamics (CFD) methods combined with Discrete Element Method (DEM) simulations have emerged as a powerful tool for studying these systems, enabling researchers to gain insights into complex particle-fluid interactions and to optimize system performance. In this study, we present a detailed investigation of DEM simulations of fluidized beds and slurries, focusing on the modeling and simulation of the fluid-particle interactions, as well as the prediction of bulk properties such as bed height and pressure drop.

Our work utilizes a novel approach to modeling the particle-fluid interactions, based on the use of coarse-grained particles that capture the essential physics of the system while reducing the computational cost of the simulation. We also incorporate a high-fidelity treatment of the fluid phase using Lattice-Boltzmann (LB)-based CFD techniques. The resulting DEM-CFD framework allows us to accurately capture the dynamics of fluid-particle systems over a range of operating conditions. We demonstrate the validity of our approach by comparing our simulation results with experimental data from packed beds, pipeline transport, and agitated tanks.

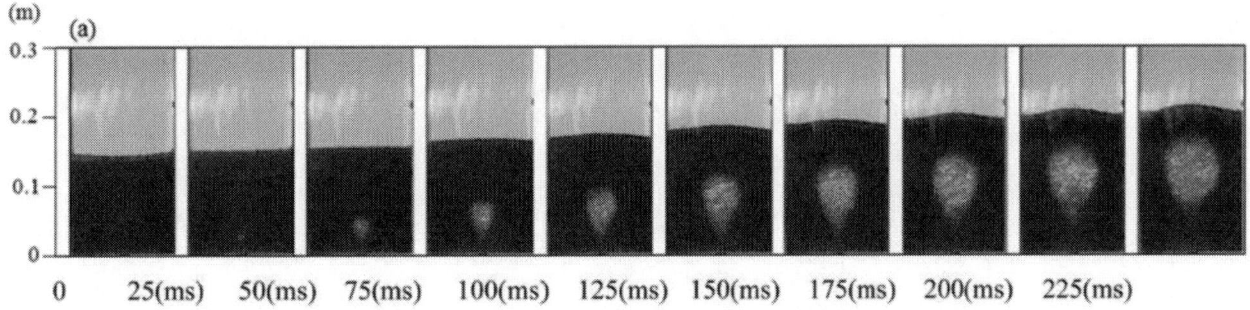

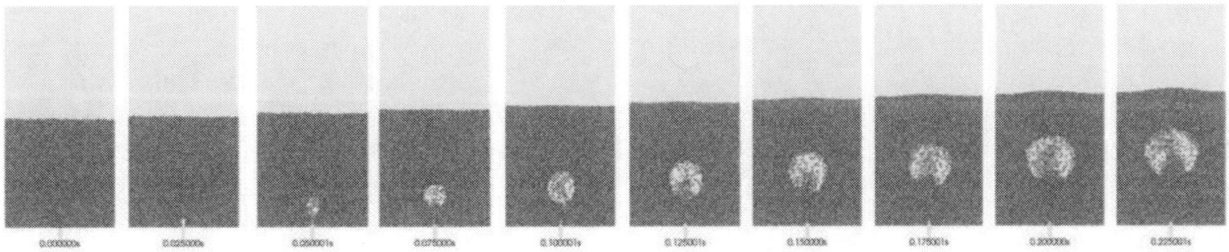

30s h^6 Organometallic Hexagonal Boron Nitride: Functionalization and Prope

Kartikey Sharma

Hexagonal boron nitride (h-BN), renowned for its formidable attributes—tensile strength (~100 GPa), wide bandgap (~6 eV), high thermal conductivity (227-280 $W.m^{-1}.K^{-1}$), and chemical stability—poses a challenge in functionalization due to its electronegativity variance between boron and nitrogen. In a breakthrough, we present unprecedented organometallic functionalization via chromium carbonyl vapor exposure. This novel approach forms a pristine η^6 bonding that preserves lattice planarity and interconnectivity between B- and N-centers, with the metal engaging each hBN ring. Computational analysis validates spontaneous surface reaction ($\Delta G = -35.50$ kcal/mol). Further, carbonyl groups seed silver nanoparticle growth, culminating in a conductive layer atop hBN. This non-destructive, chemically versatile functionalization ushers in a new era in multifunctional coatings, harnessing hBN's thermal and structural prowess across diverse applications.

73a Polymer-Grafted Plasmonic Nanocomposites with Enhanced Chiroptical Properties

Shema Rachel Abraham, Jojo P. Joseph, Alexander Baev, Paras N. Prasad, Mark Swihart

We are interested in studying nanocomposites with a higher chiroptical activity in the visible region, which can be employed in developing chiral metamaterials. We present a thin film system of chiral polymer-plasmonic nanocomposites, fabricated, and characterized to reveal enhanced chiroptical response. The polymers are synthesized using the Suzuki coupling technique with fluorene-based monomers having chiral side chains copolymerized with thiophene units to generate high optical activity in the visible region. Further, plasmonic gold nanostructures are incorporated into these polymers to achieve plasmonic enhancement of optical activity. The nanocomposites are synthesized in situ, in a solution of the polymer, to ensure the integration of the nanostructures within the helical polymer fibrils. Initially, we used monodispersed gold nanoparticles of 5nm diameter. Gold nanostructures have a strong affinity towards the thiophene units, and the *in situ* technique ensues reduced aggregation during the annealing of the thin film nanocomposites. These polymer-gold nanocomposite thin films show a 20-fold enhancement of optical rotatory dispersion compared to pristine polymer films. Further, we study the chiroptical properties of polymer nanocomposites with gold nanostars. These plasmonic multibranched nanocrystals with high-aspect-ratio spikes render a dominant localized surface plasmon resonance, thereby further enhancing chiroptical properties. The nanocomposites of chiral polymer with these anisotropic plasmonic nanostructures promote a red shift in absorbance (approximately 50 nm) and 80-fold enhancement of circular dichroism compared to pure annealed polymer film. Thus, the gigantically enhanced chiroptical response obtained here via tuning the plasmon coupling will pave the road for new metamaterials and warrant various applications in chiral polymer photonics.

73b Preceramic Polymer Grafted Nanoparticle Composites: Influence of Thermal Curing on Rheology, Microstructure, and Ceramic Yield

Gary Germanton Jr., Pitchaimari Gnanasekar, Subramanian Ramakrishnan

Preceramic polymer grafted nanoparticles (PGNPs) have unique benefits over other preceramic materials. They provide routes for increasing ceramic yield, controlling particle arrangement, and reducing volume shrinkage and mass loss. Traditionally, filler particles are introduced to free polymers, but this leads to issues of chemical compatibility, agglomeration, and phase separation when processing techniques are applied. Previous studies have established that thermal curing may be used to improve the efficiency of converting preceramic precursors (polymers and PGNPs) to the final ceramic material. Here, we investigate how the material transformations are affected by exposure to low-temperature thermal treatments prior to ceramic conversion via pyrolysis. Two PGNPs investigated in this study had a silica nanoparticle core, one having the corona poly(1,1-dimethylpropylsilnae) (allyl-GNP) and the other having poly(1,1-dimethylbenzylethylsilane) (styryl-GNP). After thermal treatment, an increase in ceramic yield was observed. This is commonly seen in commercial preceramic polycarbosilanes such as SMP-10 due to present cross-linkable sites (carbon-carbon double bond). Characterizing techniques such as rheology, X-ray photon correlation spectroscopy (XPCS), and differential scanning calorimetry (DSC) were used to elucidate the thermally included material changes in the PGNPs (Figure 1). It was determined that thermally induced changes in the PGNP structure are due to the chemical structure of these grafted nanoparticle chains. It was found that styryl-GNPs experience crosslinking between 180 °C and 250 °C, which improved the ceramic yield when thermally converted to ceramic at 800 °C and induced a transition from diffusive to ballistic behavior at the nanoscale. This was not seen in the allyl-GNPs. Oscillatory rheological measurements were performed during the thermal treatment to 250 °C, and a key finding was that thermal jamming occurs in the PGNPs with both styryl and allyl backbones, but a structure is formed faster from the styryl backbone than the allyl backbones. Improving an understanding of cross-linking behavior of these hybrid materials are expected to advance the application of these materials to a turbine engine, advanced friction, and heat-shielding components.

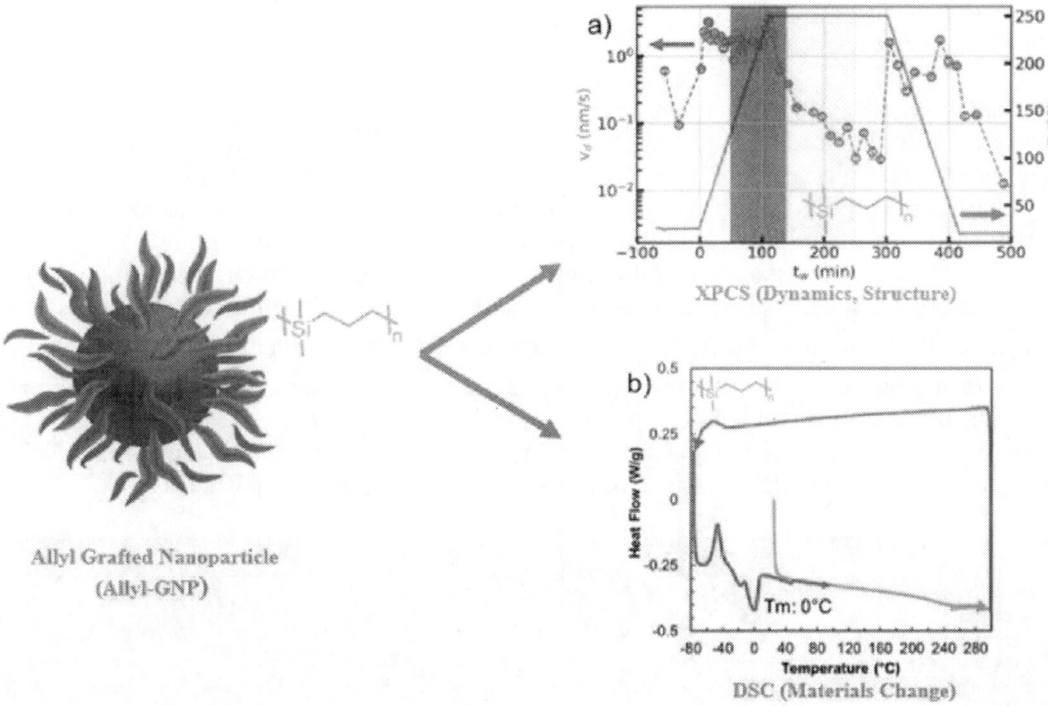

Figure 1. (a) XPCS data for allyl-GNP. Blue area represents rapid weight loss, and the yellow area represents reported weight loss according to TGA. (b) DSC for allyl-GNP. The orange line indicates the first heating, the blue line is cooling, and the red line is the second heating.

73c Perovskite-Coated Biomimetic Chiral Microparticles with Circularly Polarized Light Emission

Michael Veksler, Nicholas Kotov

A range of single-particle emitters spanning the visible spectrum and exhibiting circular polarization are formed by the coating of hybrid organic-inorganic perovskite (HOIP) onto chiral microparticles. The tunable nature of the particles allows for emissions that span the visible spectrum, while also maintaining a modest degree of circular polarization. We investigate the influence of the nanoscale and sub-microscale substrate chirality on circularly polarized light emission. This is accomplished through integrating tunable perovskite light emitters onto a substrate of hierarchically assembled biomimetic chiral microparticles of various sizes. Through a novel coating technique, we show single-particle emitters with a dissymmetry factor (g_{lum}) up to 10^{-3}. The color and dissymmetry of the emission can be controlled by the composition of the perovskite and the size/helicity of the chiral substrate, pointing to the system's potential application as tunable single-particle CPL emitters. Furthermore, the CPL intensity is not heavily influenced by the concentration of CPL emitters, showing that the CPL mechanism is dominated by scattering events at the substrate rather than through bulk or colligative properties of a collection of particles.

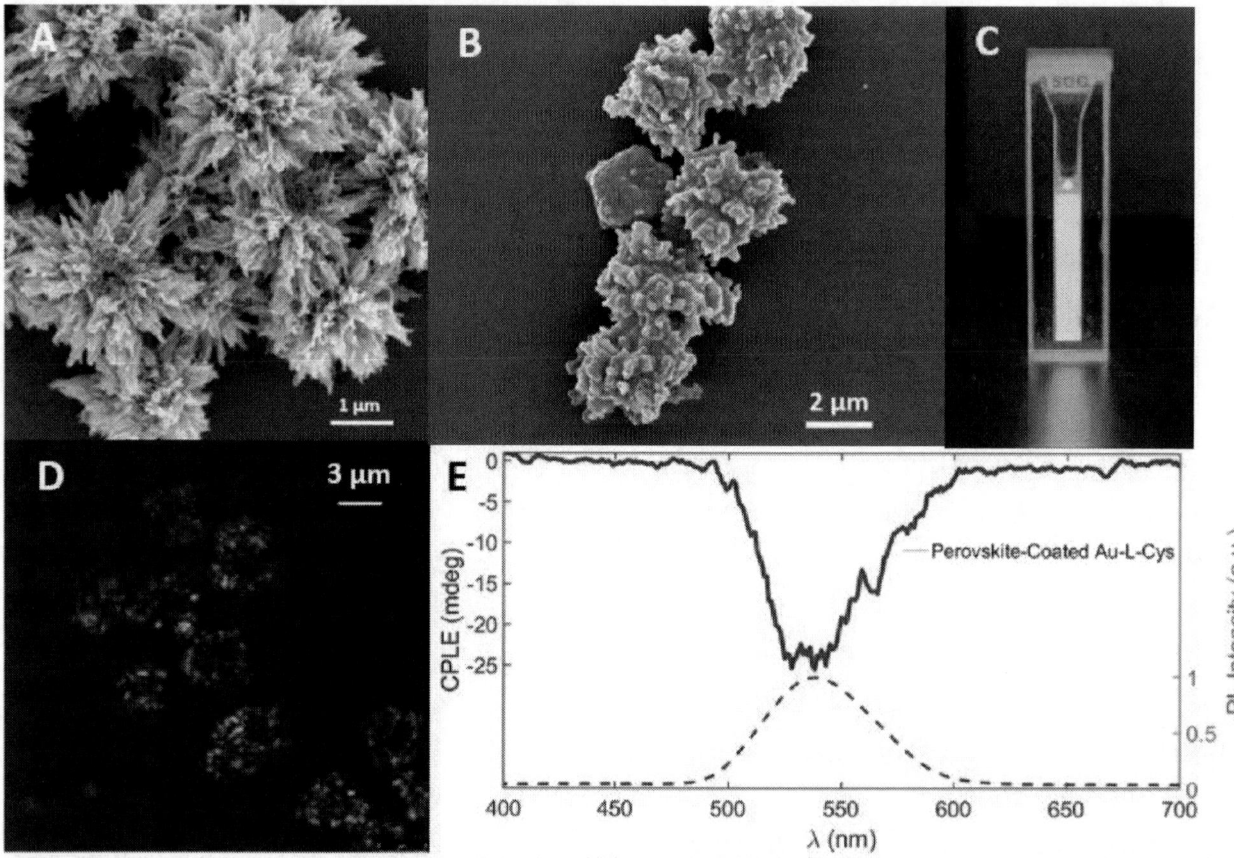

Figure. (A) SEM image Au-L-Cys particles as-synthesized. (B) SEM image of chiral Au-L-Cys particles coated with luminescent perovskite $FaPbBr_3$. (C) Photograph of colloidal dispersion of perovskite-coated chiral particles. (D) Fluorescence microscopy image of perovskite-coated particles. (E) Circularly polarized emission spectrum from perovskite-coated Au-L-Cys particles.

73e Electrically Responsive Yolk-Shell Colloidal Photonic Crystals with Controllable Coherent Scattering Intensities

Hikaru Namigata, Tom A. J. Welling, Kanako Watanabe, Keishi Suga, Arnout Imhof, Alfons van Blaaderen, Daisuke Nagao

Colloidal photonic crystals, which are the ordered assemblies of monodisperse particles, can reflect specific light wavelengths according to Bragg's law. Three-dimensionally ordered colloidal crystals can be fabricated conveniently via bottom-up methods such as self-assembly. However, in general, dynamically changing the arrangement of the composed particles in self-assembled colloidal crystals is barely possible because they form a close-packed structure. Owing to that, the scattered wavelengths and intensities cannot be tuned easily once the particles are assembled. Colloidal crystals of which composed particles can be reversibly rearranged are expected to have tunable scattering properties applicable to functional optical devices such as displays or e-ink.

In this study, yolk-shell particles (or rattle-type particles) were employed to realize the tunable coherent scattering of colloidal crystals. Yolk-shell particles consist of a hollow particle (shell) enclosing a mobile core particle. By applying an external AC electric field, the core motion changes from Brownian motion to a specific motion with reference to the field direction [1, 2]. Therefore, the core particles are expected to possess tunable mobility even if their shells form close-packed structures like often is the case when colloidal crystals are dried. We synthesized submicron-sized yolk-shell particles of the which cores are composed of silica and titanium oxide, and the shells are composed of silica. Colloidal crystals of yolk-shell particles exhibited switchable scattering intensities (i.e., Bragg reflection intensities) by applying an external AC electric field. This was due to the change in the core arrangement. Via confocal microscopy, it was observed that the core motion changed from Brownian motion to coherently moving parallel to the electric field within the shells, leading to a reduction of the displacement disorder of the cores. The coherence in the Bragg reflection could be further controlled by adjusting the electrolyte concentration in the surrounding medium [3] and/or core-to-shell size ratio, demonstrating the wide tunability in scattering intensities of yolk-shell colloidal crystals.

References:

[1] K. Watanabe *et al.*, *Langmuir*, **33**, 296–302 (2017).

[2] T. A. J. Welling *et al., J. Colloid Interface Sci.*, **627**, 761–773 (2022).

[3] T. A. J. Welling *et al., ACS Nano*, **15**, 11137–11149 (2021).

73f Essential Oil-Loaded Mesoporous Silica Nanoparticles for the Development of Multifunctional Nonfouling Coatings

Minchen Mu, Yu-Ting Lin, William DeFlorio, Yashwanth Arcot, Shuhao Liu, Wentao Zhou, Xunhao Wang, Younjin Min, Luis Cisneros-Zevallos, Mustafa Akbulut

Bacterial adhesion to the materials surfaces is one of the most common causes of bacterial infections and diseases. In the past decade, antifouling coatings have received increasing interest to tackle this grand challenge in the area of surface hygiene and microbiological safety. Herein, we report a multifunctional antibacterial coating based on simultaneous release-inactivation and superhydrophobic antiadhesion strategy. Mesoporous silica nanoparticles were synthesized and chemically functionalized to play a dual role. On one hand, deposition of silica nanoparticles generated appropriate roughness with low surface energy, which formed a superhydrophobic surface and reduced the contact between bacteria and surface. On the other hand, the particle pores were loaded with cinnamon essential oil, which is a natural bactericidal agent with health benefits to inactivate planktonic bacteria after released into the medium. Such coating was characterized by XRD, ATR-FTIR, UV-vis, SEM and TEM, and exhibited great antibacterial and antiadhesion activities with a 99.9% reduction for Gram-negative *E coli* and 99.6% for Gram-positive *S. aureus*. It also showed robust mechanical durability through scratching test and good chemical durability through acid/base treatment. Together with its advantages of facile, low-cost and universal fabrication, as well as its metal-free and fluorine-free properties, we anticipate this nanoparticle-based coating has great prospect in reducing the potential risk of bacterial contamination in biomedical applications and food industry.

73g Silica-Polyelectrolyte Nanocomposites As Stabilizers for Aqueous Microcapsules with Adjustable Properties

Matthew Lertola, Michael B. Schmithorst, Sapir Lifshiz-Simon, Michael Persson, Yeshayahu (Ishi) Talmon, Romain Bordes, Krister Holmberg, Matthew Helgeson, Bradley F. Chmelka

Particle-stabilized (Pickering) emulsions offer enhanced kinetic stability of dispersed liquid droplets, due to the adsorption of solids at the oil-water interface, preventing droplet coalescence. The simple preparation and diverse properties of these systems make them useful for applications in drug delivery, catalysis, separations, or as stiffness-enhancing agents. While numerous studies have investigated the formation of microcapsules from particle-stabilized emulsions by using inorganic particles or interfacial polymerization, the development of robust, but deformable, microcapsules for stimuli-responsive applications remains a challenging goal in soft matter research.

Here, we report the preparation and characterization of viscoelastic microcapsules formed from water-in-oil emulsions, which are stabilized by anionic silica nanoparticles, cationic polyelectrolytes, and non-ionic surfactant molecules that influence the mechanical properties of the microcapsule. The surface densities of negative charges on colloidal silica nanoparticles with different morphologies and with or without alumina surface groups were measured as functions of pH via streaming potential titrations. The respective nanoparticles were mixed with a water-soluble non-ionic surfactant and a cationic polyelectrolyte in the aqueous phase, before adding an oil (decane) phase containing an alkane-soluble non-ionic surfactant. Ultrasound-induced cavitation of the decane-water mixture resulted in the formation of submicron water-filled capsules in a continuous decane phase. Subsequent fractionation of the nanocomposite-stabilized emulsion allowed the extraction of microcapsules with a narrow distribution of sizes in the range of 600 ± 90 nm. Solid-state two-dimensional heteronuclear correlation NMR spectra of dried silica-polyelectrolyte nanocomposites revealed strong interactions between the quaternary ammonium moieties of the polyelectrolyte and the silica nanoparticle surfaces. The influences of the silica-polyelectrolyte nanocomposite on the mechanical properties of the water-decane microcapsule system was probed by interfacial rheology measurements, which revealed viscous and elastic responses that depended on the nanocomposite composition. Higher loadings of spherical silica nanoparticles, relative to elongated nanoparticles, resulted in microcapsules that exhibited greater elastic moduli across the frequency range investigated (0.1 to 200 rad/s), thereby demonstrating adjustable stiffness at the oil-water interface. Control of elasticity based on interface composition is expected to enable the design of pliable microcapsules for packaging stimuli-responsive guest species, such as pharmaceutical agents for drug delivery or protein molecules with triggerable functions.

123a Platinum Nanoparticles Efficacy and Toxicity As Cancer Treatment

Aida López Ruiz, Ashish Kokkula, Kathleen McEnnis

Nanotechnology is a multidisciplinary field that has many different applications including cancer treatment and detection. Noble metal nanoparticles such as gold, silver or palladium, have shown major potential in the field of medicine and pharmaceutics. In particular, platinum nanoparticles (Pt NPs) have been reported with promising anticancer results. The activity of nanoparticles is based on their small size and high surface area; these properties allow them to penetrate biofilms as well as influence intracellular mechanisms. Recent studies indicate that Pt NPs can be used as a therapy for cancer showing limited toxicity to healthy cells. Hence, in this work, we developed a new treatment option for triple negative breast cancer (TNBC) based on Pt NPs. TNBC is a type of breast cancer that lacks the three common receptors used for immunotherapy and with a tendency to metastasize faster. Therefore, TNBC has a worst prognosis and often chemotherapy is the only treatment option. However, chemotherapy is toxic, has many side effects and some cancers can even develop resistance, therefore, new treatment options are needed to combat TNBC.

The mechanism of action of Pt NPs is not fully understood and needs to be determined to comprehend their efficacy and toxicity as potential cancer treatment. Preliminary studies show that Pt NPs follow a similar mechanism to the platinum-based drug cisplatin where the Pt NPs enter the cell, ionize in the cell, and bind to the DNA, breaking it and inhibiting cell proliferation. Platinum based therapeutics have been a common course of treatment for TNBC. However, their toxicity and side effects prevent the patients to follow a whole course of treatment leaving them without any options. Pt NPs provide an alternative with limited toxicity for such sort of patients. However, the behavior of Pt NPs has not been studied enough *in vivo*. Furthermore, a common problem of nanoparticles is their accumulation in liver and spleen; therefore, it is important to characterize their biodistribution, before it can be used in clinical trials.

In the present study, Pt NPs were synthesized, and the anticancer activity was tested *in vitro* against TNBC, showing a remarkable cell death. As the major drawback of traditional chemotherapy is the toxicity, cell viability experiments were also performed on fibroblasts. The results showed no apparent toxicity of Pt NPs for healthy cells in contrast to cisplatin (a common chemotherapeutic drug). Based on the promising results achieved *in vitro*, the effect of Pt NPs was tested *in vivo* with mice, showing not apparent toxicity and no mice dead. Biodistribution was quantified using ICP-MS for the different organs of the mice finding the most accumulation in liver and spleen. In summary, this work provides evidence of a potential new treatment for TNBC based on Pt NPs with no apparent toxicity for healthy cells, overcoming one of the major drawbacks of current treatments.

123b Novel Polymer Modification on Spion Coatings for Improved Colloidal Stability and Surface Chemistry Control for Use in Magnetic Particle Imaging (MPI)

Hayden Good, Neel Eswaran, Ambar Velazquez Albino, Eric Imhoff, Carlos Rinaldi-Ramos

Introduction: Monodisperse Fe_3O_4 superparamagnetic iron oxide nanoparticles (SPIONs) have gathered extreme interest in use as tracers for Magnetic Particle Imaging (MPI). For these nanoparticles to be useful in biomedical applications, a polymer coating surrounding the particles is necessary to provide steric repulsion and render them colloidally stable in biological media, including cell culture media and blood. Additionally, this polymer shell offers the ability to functionalize the surface of the nanoparticles with agents that can improve their ability to become internalized by cells for labeling purposes, or to attach agents that repel cells for improved circulation half-life by reducing interaction with clearance cells. For this functionalization, the polymer shell must be clearly defined to produce nanoparticles with consistent hydrodynamic and colloidal properties. One attractive route of SPION modification is ligand exchange using silane-functional polyethylene glycol (PEG) coatings (PEG-silane), where the silane group affords strong interactions with the SPION surface. There are several ways to obtain PEG-silanes. One route is oxidation of PEG terminal hydroxyl groups using Jones reagent, followed by reaction with amine-functional silanes. However, Jones oxidation has extreme safety concerns, can lead to inconsistent polymer oxidation and can affect PEG molecular weight due to uncontrollable cleavage of the polymer backbone. Thus, this project aims to eliminate Jones reagent from a safety point of view, in addition to improving understanding of the various steps in the ligand exchange process, and their influence on the size distribution and stability of the resulting particles using a well-defined polymer precursor.

Methods: Methyl ether poly(ethylene glycol) (mPEG) was modified to bear terminal carboxylic acids opposite the methyl terminal. Two oxidation methods were utilized: Jones reagent oxidation, using strong acids and oxidizers, and a ring-opening-procedure (ROP) to open and place a succinic anhydride on the polymer. Both methods result in carboxylic acid (COOH) bearing linear PEG with a methyl group opposite the acid. However, results show that during the Jones oxidation process, the strong acids and oxidizers unpredictably splice the polymer chain, leading to smaller polymer chains that now bear two COOH groups. On the other hand, the ROP process leads to consistent molecular weight chains with well-defined chemistry. To achieve this ROP oxidation, succinic anhydride was added to dichloromethane with PEG, using EDC/NHS chemistry and 4-dimethylaminopyridine (4-DMAP) as catalysts for 24 hours, then washed and precipitated with ethyl ether. The polymer was characterized by standard ^1H NMR and Heteronuclear Multiple Bond Correlation (HMBC) NMR to confirm successful oxidation and characterized by gel permeation chromatography (GPC) to confirm polymer molecular weight was not impacted by the oxidation process. These polymers were then reacted with 3-aminopropyltriethoxysilane (APTES), an amine bearing silane group, which serves as the anchoring group of the polymer to the particle. These amine-reacted polymers were introduced to the SPIONs under high temperature to exchange the oleic acid shell in the as-synthesized nanoparticles with a siloxane-grafted mPEG shell. This reaction is done in excess of free APTES to provide increased graft density of the siloxane groups near the particle surface, while also providing free amines near the surface of the nanoparticle for additional modification. These particles were washed of free polymer and suspended in water, completing the phase transfer to aqueous solutions. The particles were characterized by DLS to determine their hydrodynamic size and colloidal stability.

Results: Polymer modifications using Jones Oxidation and a novel ROP process have been developed and demonstrated. NMR and GPC measurements show that the Jones oxidation process leads to cleaving of the polymer chain and 'over-oxidation' due to excess oxidation sites after polymer cleavage. As shown in Figure 1a, increased molar ratios of Jones reagent led to polymer peak molecular weight decrease. Not shown, the NMR characterization for these polymers showed up to 150% oxidation, compared to the methyl group present. These results confirm that this route led to polymer scission and additional sites for oxidation. On the other hand, NMR and GPC measurements show that the ROP process leads to consistent polymer length and expected oxidation rates. Figure 1b shows that as increasing molar ratios of ROP oxidation reagents were added, the peak molecular weight of the polymer did not change. Additionally, NMR of these polymer showed between 90-100% oxidation, confirming that this route led to successful oxidation of the polymer without impacting molecular weight and providing predictable polymer chemistry modifications. These polymers were reacted with APTES to link the anchoring group to the polymer, confirmed by NMR measurements while GPC measurements showed that the polymers do not agglomerate during the process. Successful ligand exchange was demonstrated resulting in particles with hydrodynamic diameters near 60 nm. Additionally, this coating was confirmed by FTIR measurements indicative of siloxane chain formation and a successful phase transfer to water. Ongoing work is focused on size and surface chemistry control and improved colloidal stability by optimizing process conditions, such as reaction solvent, temperature of reaction, molecular weight of polymer, ratios of polymer and free APTES to particles, and forms of agitation.

Conclusions: This work demonstrates that careful consideration of polymer oxidation route is important when modifying polymers. The ability to modify PEG with a ring-opening process to produce PEG bearing carboxylic acid moieties without affecting molecular weight of the polymer was developed and demonstrated. Effective coating of the tracers in a dense polymer brush and colloidal stability through steric repulsion has been demonstrated in this work. This polymer has been tested against its Jones oxidation counterpart, and shows similar ligand exchange capabilities to SPION tracers, suggesting that this new oxidation route does not impede ligand exchange results. Future and ongoing work is focusing on the ability to control size and surface chemistry, improve colloidal stability of the particles, conjugate additional ligands to the surface of the nanoparticle, and attaching thiol-reactive, folic acid, and other functional groups to the nanoparticle surface, providing targeting agent addition to the nanoparticle surface. Finally, these functionalized nanoparticles will be cultured with cells of interest to demonstrate improved cell labeling capabilities *in vitro* and performing pharmacokinetic studies to demonstrate long circulating tracers *in vivo*.

a)

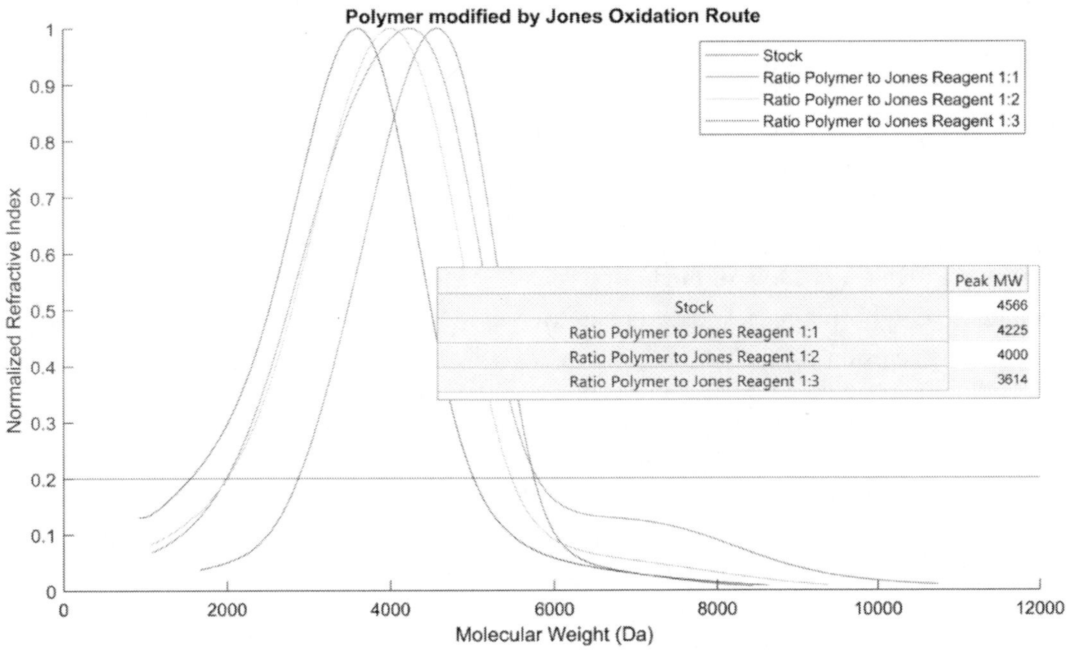

Polymer modified by Jones Oxidation Route
Stock
Ratio Polymer to Jones Reagent 1:1
Ratio Polymer to Jones Reagent 1:2
Ratio Polymer to Jones Reagent 1:3
Peak MW
Stock 4566
Ratio Polymer to Jones Reagent 1:1 4225
Ratio Polymer to Jones Reagent 1:2 4000
Ratio Polymer to Jones Reagent 1:3 3614
Normalized Refractive Index
Molecular Weight (Da)

b)

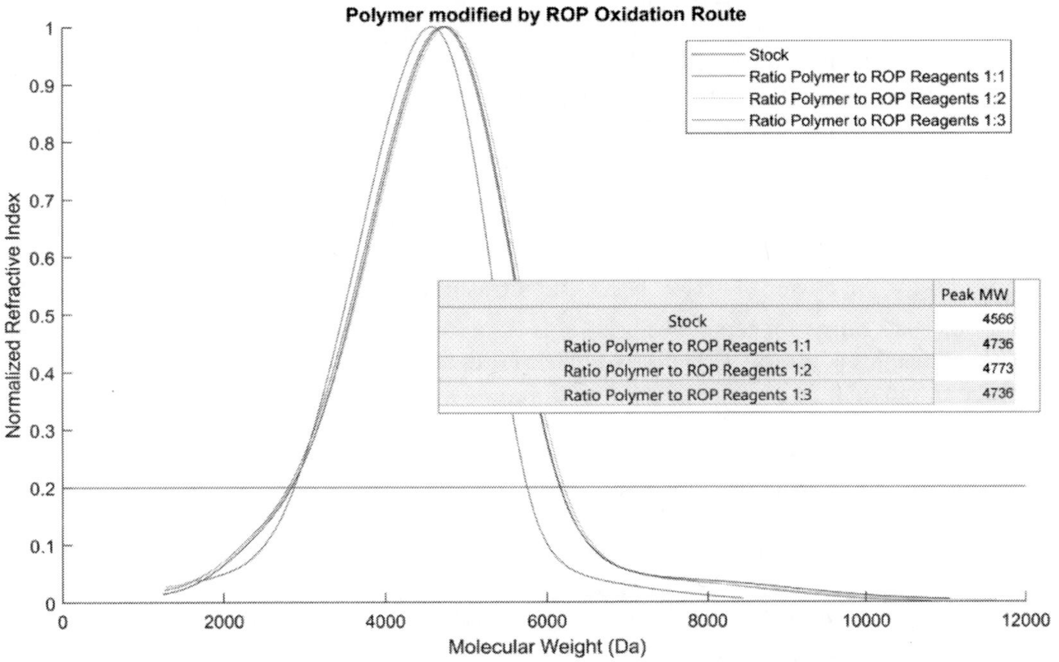

Polymer modified by ROP Oxidation Route
Stock
Ratio Polymer to ROP Reagents 1:1
Ratio Polymer to ROP Reagents 1:2
Ratio Polymer to ROP Reagents 1:3
Peak MW
Stock 4566
Ratio Polymer to ROP Reagents 1:1 4736
Ratio Polymer to ROP Reagents 1:2 4773
Ratio Polymer to ROP Reagents 1:3 4736
Normalized Refractive Index
Molecular Weight (Da)

123d Continuous Synthesis of Drug-Loaded PLGA Nanoparticles Enabled By Ultrasonic Microreactor

Aniket Udepurkar, Christian Clasen, Victor Sebastian, Simon Kuhn

Poly (D, L lactic-co-glycolic acid) (PLGA) nanoparticles have shown immense potential as carriers for drugs, proteins, and other macromolecules for their controlled release[1]. PLGA nanoparticles with a diameter in the range of 20-200 nm are best suited for drug delivery[2]. Nanoprecipitation has been implemented utilizing microreactors for the synthesis of PLGA nanoparticles (diameter 40-200 nm), however, a major drawback is a low throughput (typically below 10 mg/hr.) and class 2 toxic solvents like acetonitrile and dimethylformamide [3–5]. European Medicines Agency recommends avoiding or limiting the use of class 2 solvents due to their significant toxicities[5]. An alternative is the emulsion-solvent evaporation technique coupled with batch ultrasonic emulsification for the synthesis of PLGA nanoparticles. Emulsion-solvent evaporation utilizes ethyl acetate as a solvent, which is classified as a class 3 (low toxic potential and recommended)[5]. However, batch synthesis suffers from drawbacks such as batch-to-batch variation, lack of particle size control, the possibility of contamination, and mean hydrodynamic diameter (MHD) larger than 120 nm [6,7]. In this work, we will present an ultrasonic microreactor for the synthesis of drug-loaded PLGA nanoparticles in the desired size range (20-200 nm) utilizing emulsion-solvent evaporation technique.

The ultrasonic microreactor, consisting of a borosilicate glass reactor with a serpentine channel of square cross-section 1.2x1.2 mm^2 and a length of 700 mm (volume ~ 1 ml) bonded to a piezoelectric plate transducer of thickness 4 mm, is operated at the frequency of 550 kHz. For the generation of the PLGA nanoparticles, the continuous phase (Milli-Q water + Poloxamer 407) and the dispersed phase (PLGA (+ drug cyclosporin A) + ethyl acetate) are supplied to the reactor for the flow rate of 200 μl/min and 50 μl/min respectively (residence time 4 min). The particle size of the PLGA nanoparticles is analyzed by dynamic light scattering (DLS).

Blank PLGA nanoparticles of MHD 85 nm and polydispersity index (PDI) 0.14 are synthesized by batch solvent evaporation, for the power of 15 W. Next, the ultrasonic microreactor is coupled with a 3D-printed open channel (200x10x5 mm^3) for the continuous solvent evaporation to enable the continuous synthesis of the PLGA nanoparticles (Figure 1 (a) and (b)). The outlet of the open channel is connected to a sample collection syringe mounted on a syringe pump to maintain a steady flow. Three independent experiments for the continuous synthesis of blank PLGA nanoparticles at the power of 15 W show the results are reproducible (Figure 1(c)). In addition, no significant influence of the ultrasonic power is seen on the PLGA nanoparticle size (Figure 1(d)).

In this study, we will present an ultrasonic microreactor coupled with a 3D-printed channel for the continuous synthesis of drug-loaded PLGA nanoparticles for the desired size range of 20-200 nm. We will encapsulate the immunosuppressant drug Cyclosporin A in the PLGA nanoparticles and quantify the encapsulation efficiency and drug loading. Finally, we will study the in-vitro release of the drug from the PLGA nanoparticles. The current study will demonstrate the synthesis of monodisperse drug-loaded PLGA nanoparticles (MHD < 100 nm, PDI < 0.3) for drug delivery utilizing a non-toxic solvent.

1. Danhier, F. *et al.* PLGA-based nanoparticles: An overview of biomedical applications. *J. Control. Release* **161**, 505–522 (2012).
2. Banik, B. L., Fattahi, P. & Brown, J. L. Polymeric nanoparticles: The future of nanomedicine. *Wiley Interdiscip. Rev. Nanomedicine Nanobiotechnology* **8**, 271–299 (2016).
3. Karnik, R. *et al.* Microfluidic platform for controlled synthesis of polymeric nanoparticles. *Nano Lett.* **8**, 2906–2912 (2008).
4. Rasouli, M. R. & Tabrizian, M. An ultra-rapid acoustic micromixer for synthesis of organic nanoparticles. *Lab Chip* **19**, 3316–3325 (2019).

5. Agency, E. M. ICH guideline Q3C (R8) on impurities: guideline for residual solvents. **31**, 1–51 (2022).
6. Ding, D. & Zhu, Q. Recent advances of PLGA micro/nanoparticles for the delivery of biomacromolecular therapeutics. *Mater. Sci. Eng. C* **92**, 1041–1060 (2018).
7. Schappo, F. B., Assunção, L. S., Ribeiro, C. D. F. & Nunes, I. L. PLGA nanoparticles: Current uses and emerging trends. *Poly(Lactic-Co-glycolic Acid) Nanoparticles Drug Deliv.* 477–506 (2023) doi:10.1016/b978-0-323-91215-0.00002-9.

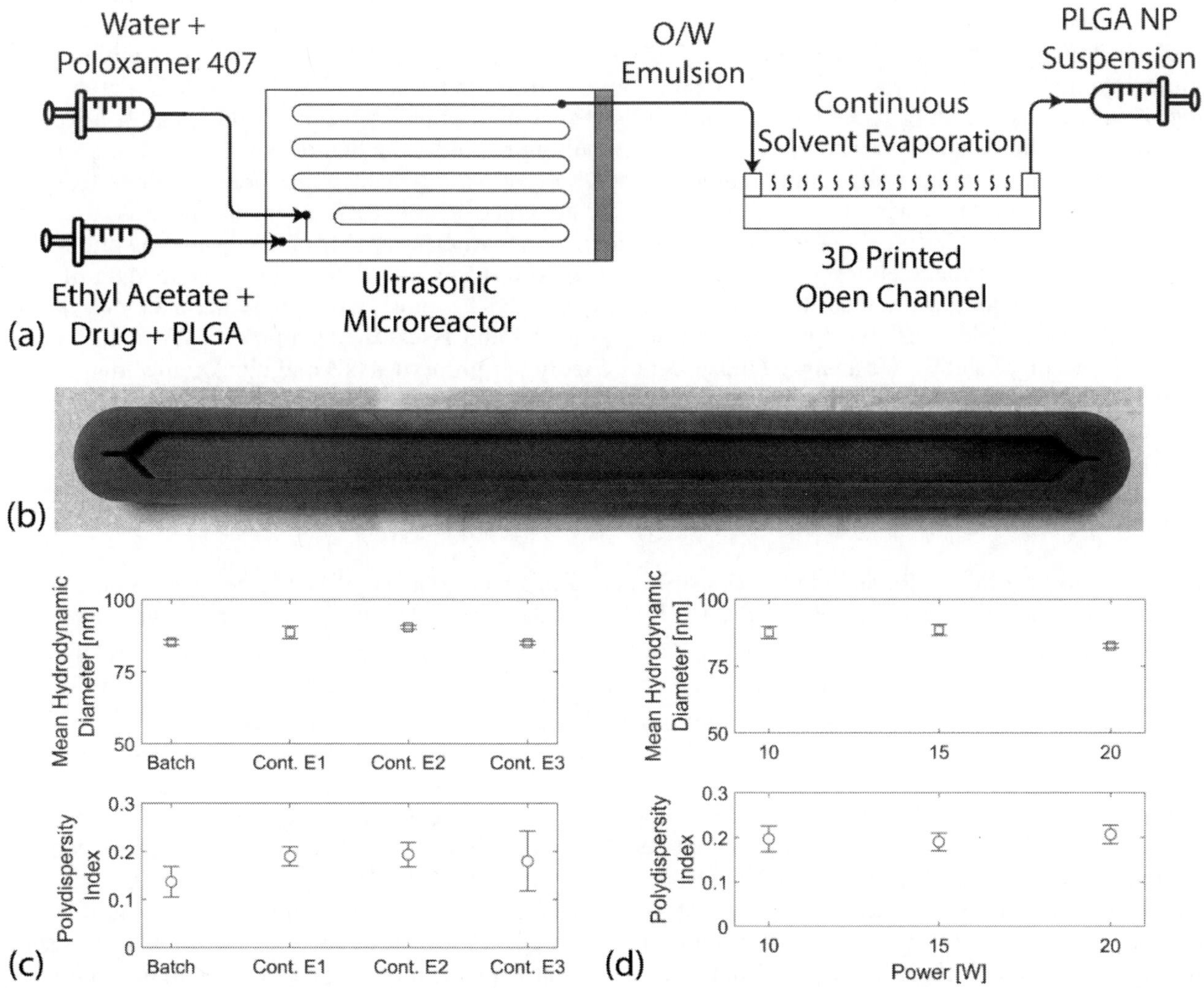

Figure 1: (a) Schematic representation of the continuous synthesis of PLGA nanoparticles, (b) 3D-printed open channel for continuous solvent evaporation, (c) Mean hydrodynamic diameter (MHD) and polydispersity index (PDI) of the three independent continuous PLGA nanoparticle synthesis (Ultrasound Frequency: 550 kHz, Power: 15W), (f) MHD and PDI of the blank PLGA nanoparticles synthesized by continuous solvent evaporation (Ultrasound Frequency: 550 kHz).

123e Endosomolytic Polymersomes Enhance Intracellular Delivery of Nucleic Acid Therapeutics for Improved Anticancer Immune Responses

Payton Stone, Hayden Pagendarm, Jessalyn Baljon, Lauren Hubert, Mina Aziz, Blaise Kimmel, John T. Wilson

Introduction: Cancer immunotherapy has revolutionized the fields of oncology and drug delivery, with immune checkpoint blockade (ICB) demonstrating remarkable potential as an alternative to traditional treatment regimens such as surgery, chemotherapy, and radiation therapy. Although ICB has demonstrated disease control in certain cancers, it has also indicated limited therapeutic efficacy in poorly immunogenic cancers. Through the activation of specific cellular pattern recognition receptors (PRRs), the immunosuppressive tumor microenvironment (TME) of these cancers can be reprogrammed to a more immunogenic, 'hot' phenotype with larger populations of infiltrating T cells, directly correlating with improved responses to ICB. Indeed, activation of the retinoic acid-inducible gene I (RIG-I) pathway has been found to elicit a downstream signaling cascade resulting in the production of type I interferons which can induce this shift in the TME. Unfortunately, most RIG-I-activating therapeutics are not able to freely cross the cell membrane where they are required to activate this cytosolic PRR. To overcome this challenge, our lab has designed pH-responsive polymeric nanoparticles capable of encapsulating and delivering anticancer therapeutics to the tumor site. The **goal** of this work is to optimize the loading of 5'-triphosphorylated double-stranded RNA (3pRNA), a potent RIG-I agonist, within polymeric nanocarriers via a facile and scalable flash nanoprecipitation (FNP) process and assess therapeutic efficacy in a cancer model. To accomplish this, we analyzed the effect of copolymer properties (second block molecular weight, alkyl chain length) on nanocarrier physical properties (size, polydispersity, surface charge, morphology), RNA loading, cytotoxicity, endosomolytic activity, and RIG-I activation *in vitro*. We then validated the ability of the system to mitigate disease progression in a murine cancer model by systemically administering our 'lead' formulation and monitoring tumor volume, weight loss, and survival over the course of the experiment.

Materials and Methods: Reversible addition-fragmentation chain transfer (RAFT) polymerization was utilized to synthesize a library of $[PEG]_{2kDa}$-*bl*-$[DMAEMA_{50\%}$-co-$A_nMA_{50\%}]_x$ copolymers with a PEG first block molecular weight (MW) of 2kDa, a DMAEMA pH-responsive component, and a second block MW of 'x'kDa where 'A_nMA' refers to an alkyl methacrylate monomer of chain length 'n' ranging from n = 2-12 carbon atoms. By inducing turbulent mixing within a confined impingement jet (CIJ) mixer of an organic copolymer solution and an aqueous solution containing solubilized 3pRNA, uniform 3pRNA-loaded nanocarriers were produced. Carriers were analyzed for size, polydispersity (PDI), and surface charge using a Malvern Nano ZetaSizer, morphology using cryogenic electron microscopy (cryoEM), and encapsulation efficiency using a Quanti-it RiboGreen assay (Thermo Fisher Scientific). Dose-dependent cytotoxicity was assessed *in vitro* using a CellTiter-Glo luminescent cell viability assay (Promega), and endosomolytic activity was measured using a galectin (Gal) reporter assay. To examine activation of RIG-I *in vitro*, reporter cells were treated with formulations and a QUANTI-Luc assay was performed to measure the amount of Lucia luciferase secreted from cells, a readout which is directly proportional to interferon production. Finally, therapeutic efficacy was assessed in a murine EMT6 breast cancer model by monitoring tumor volume, weight loss, and survival after systemic administration of 3pRNA-loaded nanocarriers.

Results and Discussion: Upon turbulent mixing and spontaneous self-assembly, drug-loaded nanocarriers were collected in an aqueous reservoir where they were then characterized as previously described. A schematic of the FNP process, experimental design, and a cryoEM image of nanocarriers are displayed in **Fig. 1**. Preliminary analysis indicated uniform nanocarriers with variable morphologies depending on second block MW and alkyl chain length. After conducting a thorough screening of particle properties using empty carriers, a potential 'lead' nanocarrier was chosen to proceed with further testing. Preliminary data indicated that the 'lead' nanocarrier with a DMAEMA pH-responsive component allowed for higher encapsulation efficiency (EE) and induced further RIG-I activation *in vitro* compared to its nanocarrier counterpart with a DEAEMA pH-responsive component, previously characterized in our lab.[1] Interestingly, increasing the inlet concentration of 3pRNA appeared to decrease EE for both nanocarriers (**Fig. 1**). In a similar manner, an initial screening with our galectin reporter system indicated increased endosomolytic activity for DMAEMA-based nanocarriers compared to DEAEMA-based nanocarriers. Future work will seek to determine polymer properties and FNP process conditions (aqueous buffer, impingement number, etc.) for 3pRNA-loaded nanocarriers of optimal physical properties capable of inducing potent RIG-I activation. We expect the optimized formulation to significantly increase RIG-I activation, mitigate tumor progression, and prolong survival in mice compared to free 3pRNA and PBS controls.

Conclusions: In this work, we report the fabrication of a library of RNA-loaded polymeric nanocarriers via a simple and versatile formulation method capable of enhancing cytosolic delivery of drug cargo and inducing potent RIG-I activation. With this FNP platform, we highlight the potential for tailored production of a variety of polymeric nanocarriers for potential translatable use in other drug delivery and immunomodulatory applications.

References: [1] Shae, D. *et al.* Endosomolytic polymersomes increase the activity of cyclic dinucleotide STING agonists to enhance cancer immunotherapy. *Nat Nanotechnol* **14**, 269–278 (2019).

Figure 1. Flash nanoprecipitation (FNP) allows for the self-assembly of 3pRNA-loaded polymeric nanocarriers under turbulent mixing conditions. After mixing, nanocarriers are collected in a scintillation vial where they can then be characterized. It was determined that the DMAEMA-based copolymer utilized in this work was able to significantly encapsulate more 3pRNA within carriers compared to the DEAEMA-based copolymer studied in our previous work.[1]

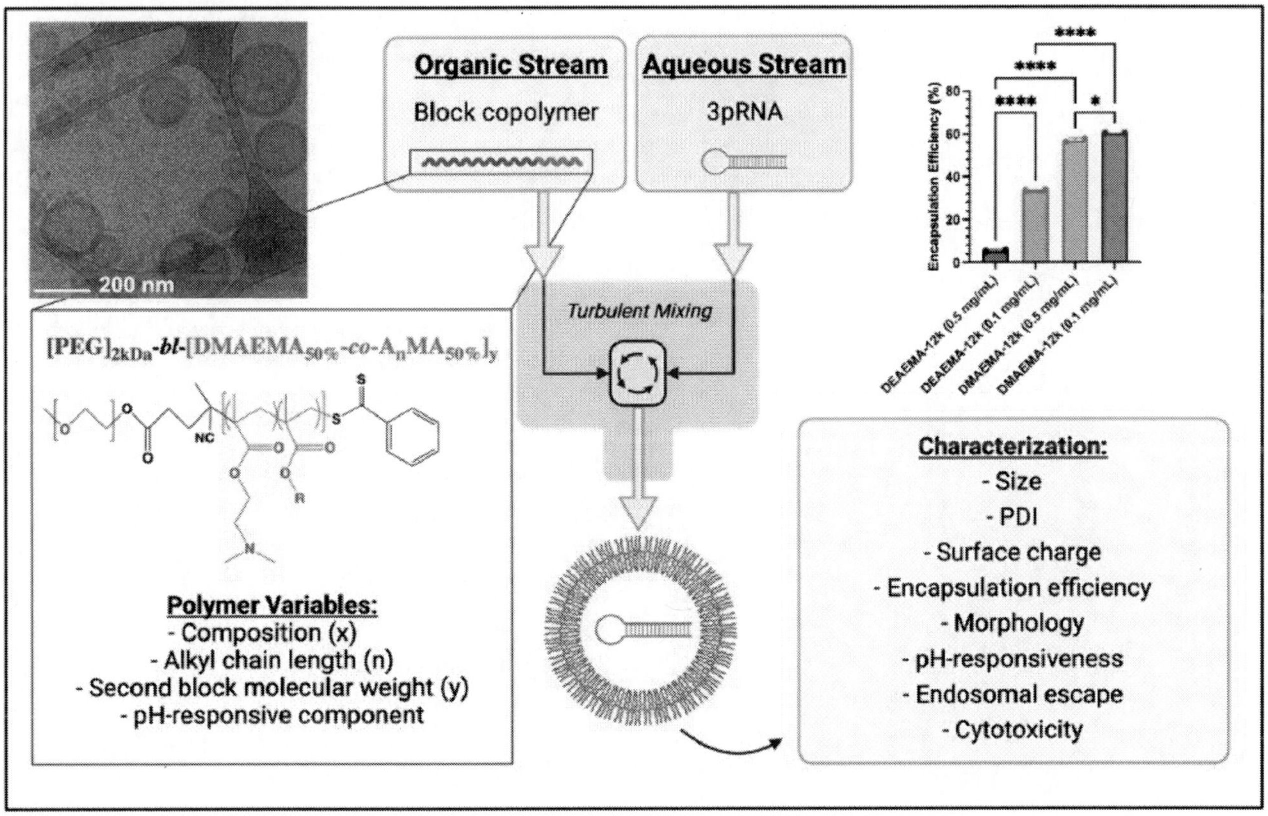

123g Antipathogenic Activities of Nitrogen, Sulfur, Boron Doped Carbon Dots

Selin S. Sagbas, Mehtap Sahiner, Alper Akcali, Ramesh S. Ayyala, Nurettin Sahiner

Nitrogen, sulfur, and boron-doped carbon dots (CDs) were prepared via the hydrothermal treatment of maleic acid (MA) as a carbon source with polyvinylamine (PVAm) as a nitrogen (N) source with cysteine (Cys) as a sulfur (S) source, and with boric acid as a boron (B) source. Different combinations such as N-, S-, B-, N/S-, N/B, N/S/B-doped CDs were synthesized and their antipathogenic capabilities on various microorganisms were examined. The blood compatibility by means of the hemolysis and blood clotting tests and the cytotoxicity on L929 fibroblast cells were determined to assess the biocompatibility of the prepared CDs. Finally, the antibacterial activities of each composition e.g., N-, S-, B-, N/S-, N/B, N/S/B-doped CDs were investigated against *B. subtilis* ATCC 6633 (Gram +), *Staphylococcus aureus* ATCC 6538 (Gram +), *Escherichia coli* ATCC 8739 (Gram -), *P. aeruginosa* ATCC 10145 (Gram -) and *Klebsiella pneumoniae* ATCC 700603 (Gram -) bacteria species via disc diffusion and microtiter dilution assays. Furthermore, biofilm inhibition and disruption ability of the most effective CDs were determined against *Candida albicans* ATCC 10231.

Acknowledgments: This work is supported by Scientific Research Commission of Canakkale Onsekiz Mart University FIA-2022-3667.

182a Scalable Manufacturing of Nanostructured Materials for Energy Applications Using Gas Phase Deposition

J Ruud Van Ommen

Nanostructured solids (for example, core-shell nanoparticles or porous micro-particles containing nanoparticles) hold great potential for energy applications, including catalysis and energy storage. However, to come from product design incorporating nanostructuring to actually making such materials at relevant scales is not trivial. Gas phase deposition using techniques such as chemical vapour deposition (CVD), atomic layer deposition (ALD), and molecular layer deposition (MLD) can be used to provide the surface of a particle with either an ultrathin continuous coating or a decoration of nanoclusters. This allows us to take benefit from the vast range of chemistries that has been developed over the years for CVD, ALD, and MLD.

Gas phase deposition is also attractive for the potential to manufacture nanostructured particles on a large scale when implemented in a system where particles are suspended in the gas phase, such as a fluidized bed or a pneumatic transport reactor. Even nanopowders can be processed using this method, although they form very dilute agglomerates and are not fluidized as individual particles. Since we often process cohesive materials, an effort has been made to obtain proper dispersion of the particles.

In this presentation, experimental evidence will be given of the benefits of applying gas phase deposition to powders. The application to various areas of catalysis will be discussed, such as TiO_2 modified with SiO_2 and Pt for photocatalysis [1], and mixed Pt/Pd particles for electrocatalysis [2]. I will also show how gas phase deposition can be deployed to enhance the lifetime of catalysts. We conclude that combining gas phase deposition with suspending particles either in a fluidized bed reactor or in a pneumatic transport reactor is an attractive way to combine nanoscale precision with producing large amounts of catalytic materials.

[1] Benz, D., Van Bui, H., Hintzen, H. T., Kreutzer, M. T., & van Ommen, J. R. (2022). Mechanistic insight into the improved photocatalytic degradation of dyes for an ultrathin coating of SiO2 on TiO2 (P25) nanoparticles. *Chemical Engineering Journal Advances*, *10*, 100288.

[2] Li, M., Fu, S., Saedy, S., Rajendrakumar, A., Tichelaar, F. D., Kortlever, R., & van Ommen, J. R. (2022). Nanostructuring Pt-Pd Bimetallic Electrocatalysts for CO2 Reduction Using Atmospheric Pressure Atomic Layer Deposition. *ChemCatChem*, e202200949.

182b Atomic Layer Deposition Nano-Engineered Ni/Al_2O_3 Catalyst for Bi-Reforming of Methane

Xinhua Liang, Baitang Jin, Kaiying Wang, Han Yu

CeO_x-overcoated Ni/Al_2O_3 catalysts were prepared by atomic layer deposition (ALD) for bi-reforming of methane (BRM), as the combination of dry reforming of methane (DRM) and steam reforming of methane (SRM). Ni nanoparticles were deposited on alpha-Al_2O_3 nanoparticles by ALD, followed by deposition of non-stoichiometric CeO_x thin films on Ni/ Al_2O_3 particles by ALD, which constructed a beneficial Ni-CeO_x interface and modified the catalyst property. Ascribed to the unique ALD growth mode, a high amount of Ce(III) and oxygen vacancies existed in the ALD-deposited CeO_x overcoating. A reduction process before the BRM reaction contributed to the further reduction of Ce(IV) to Ce(III), resulting in more oxygen vacancies. The oxygen vacancies at the Ni-CeO_x interface enabled a high rate of CO_2 activation and enabled the balance between the activation of CO_2 and H_2O for BRM. Due to its oxygen vacancies as activation sites for CO_2 and H_2O, CeO_x ALD overcoating significantly improved the activity of Ni/Al_2O_3 catalyst and achieved a better control in the H_2/CO ratio with a suitable ratio of $H_2O/CO_2/CH_4$ feed. CeOx overcoatings enhanced the reducibility of Ni(II) sites and assisted in preventing Ni from oxidation during the BRM reaction. Less carbon deposition was achieved by the Ni/Al_2O_3 catalyst with CeO_x overcoating as ascribed to its better reactant activation capacity.

182c Synthesis of Accessible and Catalytically Active Sites on Sub-Nanometer Ru Nanoparticles for CO Oxidation and H_2O_2 decomposition Reactions

Sayed Abu Sufyan, Michael Nigra

Ruthenium has demonstrated catalytic activity for a wide variety of reactions which include the reduction of C=C and C=O bonds, ammonia synthesis, and Fischer-Tropsch synthesis as several examples. In this work, bulky organic phosphine and thiol ligands are used to synthesize sub-nanometer Ru nanoparticles. The bulky ligands provide a steric barrier against nanoparticle aggregation and enable accessibility of reactants to the nanoparticle surface through inefficient packing on the Ru surface. The particle size distribution is measured by high-angle annular dark field scanning electron microscopy (HAADF-STEM), and the average diameter of the Ru nanoparticles is approximately 0.8 nm. All ligands used in the work give particles of approximately the same size. These are believed to be the smallest ligand-stabilized Ru nanoparticles synthesized in solution in the literature.

The Ru nanoparticles exhibit catalytic activity for CO oxidation and H_2O_2 decomposition with the ligands attached to the Ru surface. All the nanoparticles exhibit the same CO oxidation rates, and the identity of the functional group (thiol versus phosphine) bound to the Ru nanoparticle does not change the CO oxidation rate. However, in H_2O_2 decomposition, the identity of the ligand does play an important role in this reaction. It is observed that the hydrophobicity/hydrophilicity of the ligand has a significant impact on the H_2O_2 decomposition rate with a more hydrophobic ligand layer leading to reduced H_2O_2 decomposition rates. Controlling H_2O_2 decomposition rates is important in reactions which use H_2O_2 as a reactant because the catalysts for these reactions are also often active for H_2O_2 decomposition. Thus, it is demonstrated that bound organic ligands provide a promising approach to achieving control of H_2O_2 decomposition rates.

182d Methanol Production By CO_2 Hydrogenation in Hybrid Membrane-Nanoparticle Reactors

Eirini Goudeli, Quang Huy Pham, Colin Scholes

Methanol presents a range of advantages as a potential future fuel due to its high energy density compared to that of natural gas and hydrogen. In addition, it is a crucial precursor for the production of plastics, fuel additives and adhesives in a range of industries. However, traditional methods of methanol synthesis by natural gas have a significant greenhouse gas footprint as they involve CO_2 release as a byproduct. An alternative approach to methanol production by natural gas, is CO_2 hydrogenation, which is key in mitigating CO_2 emissions. However, several challenges associated with this approach include low reactivity of CO_2 and high selectivity to methanol. So, research interest has focused on the development of new catalysts, such as Cu/ZrO_2 (Kattel et al., 2016), $Cu/ZnO/ZrO_2$ (Witoon et al., 2018), and $Cu/ZnO/Al_2O_3$ (Tada et al., 2017), that are both active to CO_2 hydrogenation and selective to methanol.

Here, nanoparticle-membrane nanocomposites are manufactured via direct one-step deposition. CuO/ZrO_2 nanoparticles are produced by flame spray pyrolysis (FSP) at various precursor solution-to-oxygen (P/O) flow rate ratios and are deposited on polymeric membranes that are highly selective to methanol (Figure 1a). Figure 1 shows scanning electron microscopy (SEM) images of flame-made CuO/ZrO_2 catalyst produced at a P/O of (b) 10 mL/min precursor solution to 8 L/min oxygen (10/8, hot flame) and (c) 2/8 (cold flame), deposited on Matrimid®5218 membranes for 2.5 and 9 min, respectively. Increasing the precursor flow rate, leads to higher temperature and longer high-temperature particle residence time, which in turn results in larger CuO and ZrO_2 crystal size and smaller specific surface area. The catalyst is reduced by H_2 at 300 °C for the formation of Cu/ZrO_2 nanocatalysts.

The effect of nanoparticle and film characteristics (crystallinity, crystal size, porosity) on the CO_2 conversion and methanol production rate is quantified. FSP conditions leading to nanocatalysts with highly activity to CO_2 hydrogenation are identified, while the selectivity to methanol, which is enhanced by the presence of polyimide membranes, is also evaluated. Such hybrid membrane-catalytic nanoparticle reactors are a promising technology which enables industries to convert their carbon dioxide wastes into valuable energy products, such as methanol.

Funded by Future Fuels Cooperative Research Centre grant (RP1.3-04) and the University of Melbourne, Australia.

Kattel, S., Yan, B., Yang, Y., Chen, J. G., and Liu, P. (2016). *J. Am. Chem. Soc.* 138: 12440-12450.

Witoon, T., Numpilai, T., Phongamwong, T., Donphai, W., Boonyuen, C., Warakulwit, C., Chareonpanich, M., and Limtrakul, J. (2018). *J. Chem. Eng.* 334: 1781-1791.

Tada, S., Watanabe, F., Kiyota, K., Shimoda, N., Hayashi, R., Takahashi, M., Nariyuki, A., Igarashi A., Satokawa, S. (2017). *J. Catal.* 351: 107-118.

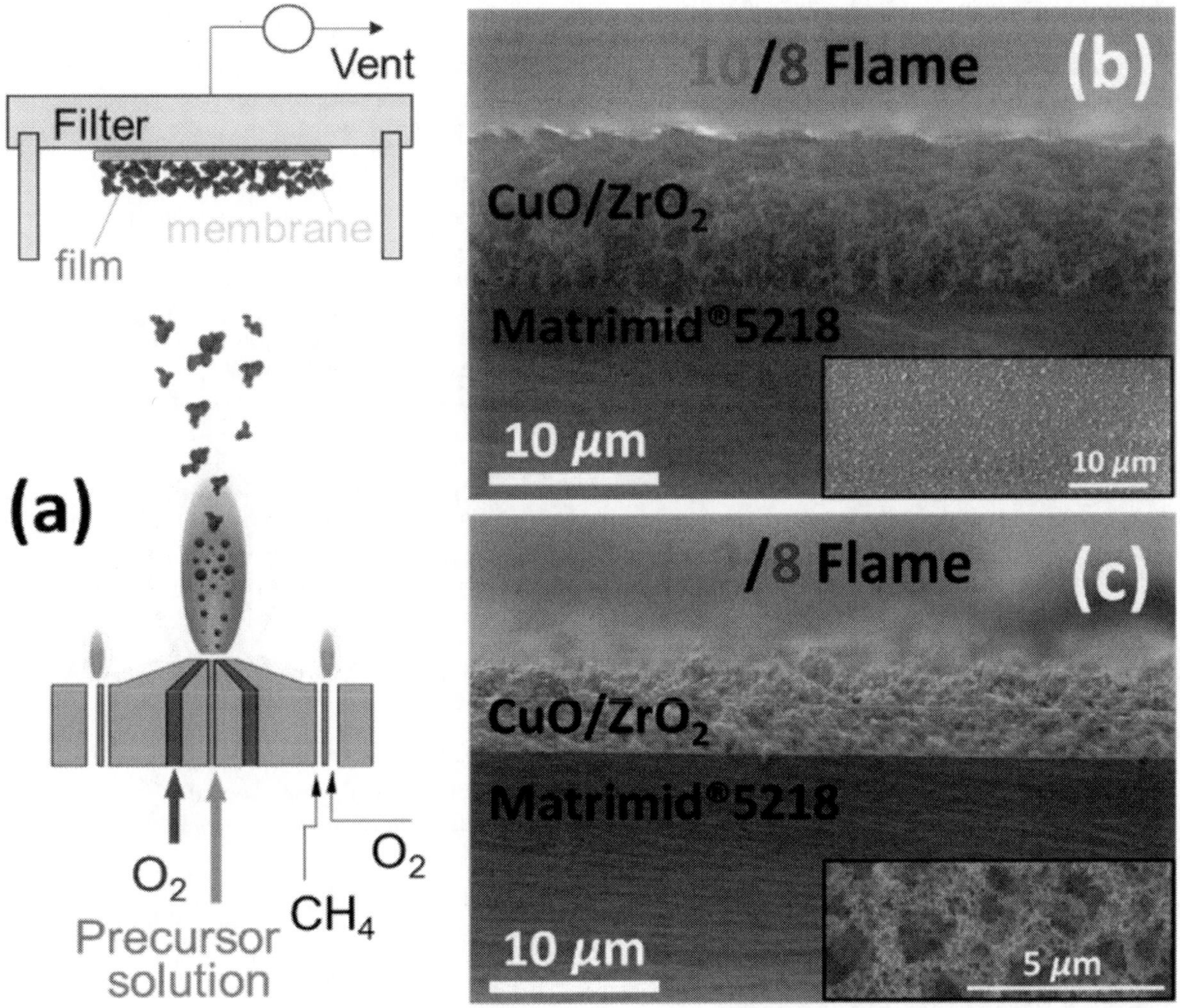

Figure 1. (a) Experimental setup for direct, one-step deposition of flame-made nanoparticles on polymeric membranes. SEM images of CuO/ZrO_2 nanoparticles produced in (b) 10/8 and (c) 2/ flames, deposited on Matrimid®5218.

182e Functional Nanoparticles and Nanocomposites for Treatment of Contaminants in Water: From Materials to Devices

Angira Das, Pritam Biswas, Rajdip Bandyopadhyaya

Taking recourse to nanoparticles and their composites for making more functionally efficient materials for sustainable solutions is a modern theme across several disciplines. We have applied this for removal of certain contaminants (e. g. microorganisms like *E. coli,* metals and antibiotics), which is a core requirement for having access to clean water.

With regard to microbial contamination, it is imperative to have a low-cost and continuous supply of potable water, devoid of any microbial population. To this end, we have developed silver and copper nanoparticle-impregnated activated carbon (see figure) as a cheap, sustainable material which have been demonstrated to produce clean water, on treating *E. coli* containing input water. We have done systematic experiments and analysis of batch, shake-flask experiments initially, and then with continuous filter-columns, to discern the mechanism of complete killing of bacteria and the resultant kinetics of water disinfection. Substantially enhanced disinfection rate was possible due to synergy between silver and copper nanoparticles.

Subsequently, this has been translated to the design of two successive generations of gravity-driven water disinfection devices achieving the target of continuous production of potable water, which runs without the use of any electricity or other power sources (see figure). For example, on taking *E. coli* of 10^4 CFU/ml as continuous source of contaminated water, we have used our device to produce disinfected water continuously up to several weeks over many hundreds of liters, suitable for a household.

Presently, ongoing experiments in our designed microfluidic device mimics the intergranular-carbon pores, simulating the nanoparticle-carbon-bacteria three-way contact, to gain fundamental insights into this biotic-abiotic interaction, driving the enhanced water disinfection rate.

Along similar lines, we investigate other nanoparticle composites developed by us, for removal of contaminants, like metals, antibiotics etc., which are critical in either drinking water or wastewater treatment applications.

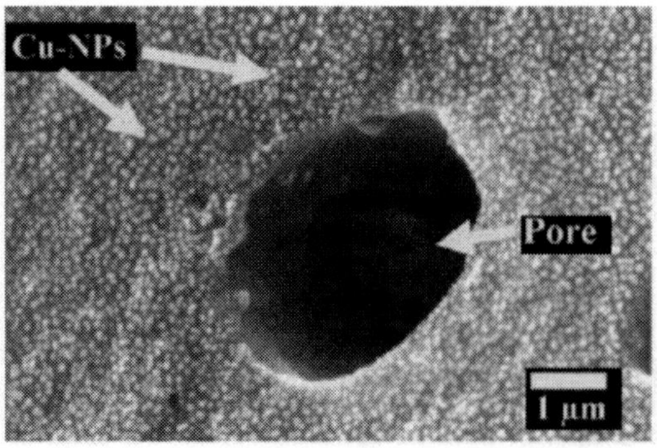

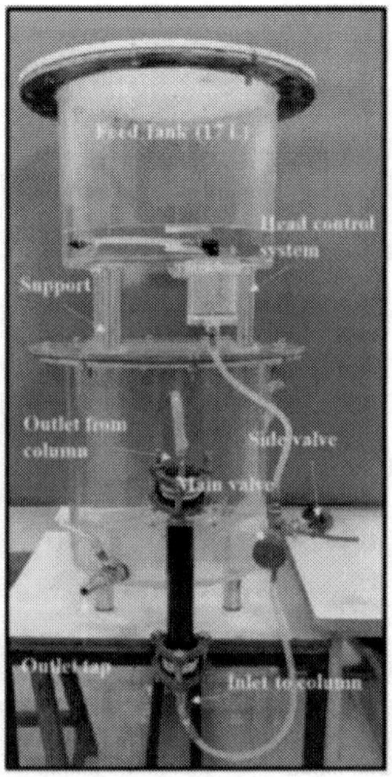

182f Fabrication of Ti_3C_2-Polyacrylate Nanocomposites By High-Temperature Free-Radical Polymerization

Vahid Rad, Hossein Riazi, Masoud Soroush

MXenes are a family of nanomaterials that are hydrophilic and electrically- and thermally conductive and have surface functional groups such as hydroxyl, oxygen, chlorine, and fluorine. Similar to graphene oxide and clay, MXenes have layered structures and possess excellent antibacterial properties and high light-to-heat conversion efficiency [1, 2].

Spontaneous high-temperature (>140°C) free-radical polymerization of alkyl acrylate monomers has several appealing features [3-5]. First, the polymerization does not require any conventional initiator. Second, it results in the production of low average-molecular-weight polymers due to the more dominating role of polymerization reactions such as beta-scission and chain transfer to monomer at high temperatures. Third, it is versatile; it can be utilized to synthesize homopolymers and copolymers from monomers with an unsaturated carbon double bond. Low average-molecular-weight polymers have low solution viscosity. The addition of a 2-dimensional (2D) nano-filler such as Ti_3C_2 MXene to a reacting monomer leads to viscosity buildup, particularly at a high filler loading. In view of these, high-temperature free-radical polymerization is an attractive method for the in-site synthesis of nanocomposites.

In this work, we study the fabrication of Ti_3C_2-poly-alkyl-acrylate nanocomposites using (a) *in-situ* high-temperature polymerization of several alkyl acrylate monomers (including *n*-butyl acrylate) in the presence of Ti_3C_2 MXene and (b) the pristine MXene is replaced by a surface-modified MXene containing methacrylate groups. The prepared nanocomposites are characterized using a Scanning electron microscope, X-ray photoelectron spectroscopy, X-ray Diffraction, and Fourier-transform infrared spectroscopy. The effects of polymerization conditions and monomer type on nanocomposite properties (such as conductivity, thermal stability, and mechanical properties) are investigated.

Keywords: Ti_3C_2 MXene, Alkyl acrylates, Free-radical polymerization

References

[1] Anasori B, Gogotsi Y. MXenes: Trends, growth, and future directions. Graphene and 2D Materials. 2022:1-5.

[2] Anasori B, Lukatskaya MR, Gogotsi Y. 2D metal carbides and nitrides (MXenes) for energy storage. Nature Reviews Materials. 2017;2(2):1-17.

[3] Riazi H, A. Shamsabadi A, Grady MC, Rappe AM, Soroush M. Experimental and theoretical study of the self-initiation reaction of methyl acrylate in free-radical polymerization. Industrial & Engineering Chemistry Research. 2018;57(2):532-9.

[4] Liu S, Chua L, Shamsabadi AA, Corcoran P, Patra A, Grady MC, et al. Oxygen-Initiated Free-Radical Polymerization of Alkyl Acrylates at High Temperatures. Macromolecules. 2021;54(17):7925-30.

[5] Riazi H, Arabi Shamsabadi A, Grady MC, Rappe AM, Soroush M. Method of moments applied to most-likely high-temperature free-radical polymerization reactions. Processes. 2019;7(10):656.

182g Energy Upgrade of a Building with TiO_2 Coated Roof Tiles Using the Energy Plus Simulation Program

Marianthi Kermenidou, Maria Kouroutzi, Antonios Stratidakis, Catherine Gabriel, Maria Perikli, Eleni Feleki, Spyros Karakitsios, Dimosthenis Sarigiannis

Introduction

The huge energy demands buildings require for cooling have led the search for innovative solutions to ensure energy performance in the summer months. One of the most promising solutions is that of "cool roofs". These have high solar reflective properties and absorb less solar energy. Materials with high reflectiveness (NIR) are in great demand in applications such as energy saving roofs, walls and pavements When an object absorbs solar radiation, energy from the sun is converted into heat energy, while approximately one third of the heat absorbed by one building originates from its roof. A study conducted by Selvaraj et al. [1] found that by increasing the heat reflection from the roof of their house the owners saved an average of 23% on house cooling costs. Another study by Akbari et al. [2] presented the results of applying a cool roof on a retail shop in Montreal, Canada, and found an annual saving of 60$ per 100m^2 roof. A simulation study by Konopacki et al. [3] concluded that having a cool roof can reduce annual energy costs in most regions of the world that require air conditioning in the summer season. In this study, the cooling loads required for the summer months by a nursery school located in Thessaloniki, Greece, were calculated for a roof of natural red clay tiles coated with TiO_2 and PEG. Simulation of the energy behavior of the building was performed using EnergyPlus™ software.

Methodology

Synthesis of TiO_2/PEG nanoparticles was conducted by adding 2.5 g of titanium dioxide into 100 mL of ionized water. The suspension was stirred and sonicated for 10 minutes at each phase of the development procedure. After this time, 180 μL of polyethylene glycol was added dropwise into the mixture and stirring was started for a total time of forty minutes with a sonication bath (20 minutes) at the half-way time point.

Natural and TiO_2 coated red tiles were characterized through solar reflectance and thermal emission measurements. Spectral reflectance measurements were carried out using the Jasco UV/VIS/NIR spectrophotometer, model V670, in the wavelength range of 250-2500 nm. Thermal emissivity measurements were carried out using a Devices and Services emissometer instrument, model AE1.

The EnergyPlus program was used for the energy simulations. Together with Open Studio, this program forms part of the usual energy modeling and simulation study of any building. Moreover, the Sketch Up program was used to construct the spaces of the energy model. This is a design program that allows the three-dimensional design of any object. The geometry of the building input into the program was based on the original design plan of the school. The three-dimensional thermal model was generated according to the geometry of the building. Each room/space was separated and building openings were included in the model. Additionally, the external conditions of each surface were defined. Following this, construction sets that correspond to the structure of each surface were created and the use of each space was defined. To estimate its energy efficiency, the building was divided into thermal zones, i.e., areas with similar uses and/or the same electromechanical systems. The complex design of the building and its variety of area usage led to the creation of six different thermal zones. These zones account for the air-conditioned spaces within the building. The roof is not air-conditioned but does form part of the building and does play a major role in its energy performance. Therefore, it was defined as an additional thermal zone. The building comprises one floor with an area of 1313 m^2 and a roof with an area of 1153 m^2. The roof is formed of the following construction elements: roof tile, roof membrane, roof insulation, roof membrane and wooden frame. Based on the above, the coated roof tile was evaluated as to how it influences indoor operative temperature, which impacts indoor thermal comfort perception.

The characteristics of the natural red clay tile used by the program to calculate thermal loads were: 0.4 W/m*K (conductivity), 2000 kg/m^3 (density), 800 J/kg*k (specific heat), 0.42 thermal absorptance, 0.61 solar absorptance and 0.77 visible absorptance, while the characteristics of the roof tile coated with TiO_2 (anatase) and PEG were: 0.4 W/m*K (conductivity), 2000 kg/m^3 (density), 800 J/kg*k (specific heat), 0.2 thermal absorptance, 0.19 solar absorptance and 0.15 visible absorptance. In this way, each change in the model would generate a corresponding change in its energy consumption and the real percentage of saved energy could be determined for each scenario. Thermal analysis of the coated roof tile was carried out in two seasons (summer and winter) to account for the climate of the particular region of the school's location.

The city of Thessaloniki is located in northern Greece within climate zone C. The school is heated from 15th October until 30th April each year, and is cooled from 1st June to 31st August each year. The total number of heating degree days differing from the optimal 20°C is estimated as 2184 and the total number cooling degree days for the limit of 26°C is estimated as 135. Hourly meteorological data site were collected from the database Europe WMO Region 6 (Thessaloniki 166220 (IWEC) and used by the EnergyPlus engine.

Results

Measurements of total solar reflectance (TSR) and UV, VIS, NIR were carried out on the natural clay tile, the clay tile coated with PEG, the clay tile coated with TiO_2 anatase, and the clay tile coated with TiO_2 anatase and PEG. The results showed that the reflectance spectrum is significantly improved in the UV, VIS and NIR regions until approx. 1900 nm and that the optical absorption of TiO_2 anatase significantly increases at 400nm and above. Regarding the tile coated with the TiO_2 anatase and PEG mixture, the TSR increased by 51% compared to the TSR of the natural red clay roof tile. Values of natural tiles and tiles coated with PEG alone were at the same level, indicating that PEG does not contribute to reflectance. Thermal emission measurements showed that tiles coated with TiO_2 (anatase) and PEG have slightly higher emission values than the natural clay tiles. Results of the simulations with Energy Plus to calculate the cooling and heating loads required per month when the roof is covered with natural clay tiles or clay tiles coated with TiO_2 and PEG, showed that the cooling loads required to maintain the temperature inside the building below 26°C in the summer, reduced significantly when the proposed coated roof tiles were applied. In July and August when the hottest environmental temperatures are recorded the cooling loads were reduced by 24% and 23%, respectively. The cooling loads were zero in the winter months. Regarding the thermal loads, there is an average increase of around 7% in the winter. The number of hours in summer with temperatures higher than 26°C was calculated within each of the building thermal zones when, a) applying natural red clay roof tiles, and b) applying roof tiles coated with TiO_2. It was found that when the roof tile coated with TiO_2 was applied, the percentage of hours with T>26°C inside the building reduced by 20-44% depending on the thermal zone.

Conclusions

The results of this study showed that there is a significant difference in cooling load consumptions when applying natural red clay roof tiles and clay roof tiles coated with TiO_2 & PEG. The total cooling load required to maintain temperatures below 26°C in the summer is 63.450 kWh when roof tiles coated with TiO_2 & PEG are applied and 82.828 kWh when uncoated natural red clay roof tiles are applied. On the contrary, the thermal load required in winter to maintain a temperature >20°C is 42.308 kWh when roof tiles coated with TiO_2 & PEG are applied and 39.203 kWh when uncoated natural red clay roof tiles are applied. Therefore, energy consumption is reduced by 13.3%. Given that the price of 1 kWh in Greece for the first half of 2022 was set at 0.2305 euro/kWh (higher than the average EU price of 0.2525 euro/kWh) [4] the percentage reduction of energy required for cooling brought about by the use of cool roofs can be economically significant for both private households and the public sector, not only in Greece but also in other countries that experience high summer temperatures.

Funding sources

This research study was funded by the Operational Programme Competitiveness, Entrepreneurship and Innovation 2014-2020 (EPAnEK) of the Ministry of Economy and Development, aiming to develop a new bioclimatic product which will have photoreflective and photocatalytic properties for use in cool roofs.

References

1. Meenakshi, P. and M. Selvaraj, *Bismuth titanate as an infrared reflective pigment for cool roof coating.* Solar Energy Materials and Solar Cells, 2018. 174: p. 530-537.
2. Hosseini, M. and H. Akbari, *Effect of cool roofs on commercial buildings energy use in cold climates.* Energy and Buildings, 2016. 114: p. 143-155.
3. Konopacki, S., *Cooling energy savings potential of light-colored roofs for residential and commercial buildings in 11 US metropolitan areas.* 1997.
4. Eurostat. *Electricity price statistics.* 2023 7 March 2023, at 16:19; Available from: https://ec.europa.eu/eurostat/statistics-explained/index.php?title=Electricity_price_statistics.

275a Estrogen-Loaded X-Ray CT Contrast Agents for Preclinical Detection of ER+ Breast Cancer

Ankan Biswas, Alice Truong, Mary Ruskowski, Manos Gkikas

Breast cancer is the most common cancer and the 5th highest cause of cancer mortality. The number is projected to increase 39% by the year 2040. Early detection of the disease is thus highly important. Mammography and X-ray-computed tomography (CT) are considered the most suitable medical imaging methods in terms of patient comfort and cost. ***Non-invasive disease detection with pronounced signal enhancement at the site of tumor is a great challenge in medicine*** and allows for condition estimation at early stages. Designing metallic nanoparticle (NP) contrast probes for spectral CT imaging that can be statistically differentiated from biological fluids, tissue, and bone (a property that is impossible to achieve with conventional CT), and are empowered by *molecular recognition*, aligns towards that direction. Our Team (UML-Chemistry/UMMS-Radiology) is developing ***an X-ray contrast probe that is specifically targeting estrogen receptor positive (ER+) breast cancer cells and tumors,*** aiming to achieve *early disease diagnosis* and *enhanced CT detection at low NP dosages.* Among the metallic NPs, Au offers significant advantages towards materials differentiation since its K-edge is 81 vs. 4 keV of Ca, while its density is 19.6 vs. 1.6 g/mL of Ca and 1.0 g/ mL of soft tissue. Taking advantage of the fact that the X-ray absorption is highly dependent on the K-edge and density of materials, ***we have synthesized hydrophilic, estradiol-PEG-AuNPs with high delivery inside ER+ breast cancer cells via estradiol-ER interactions and CT contrast comparable to that of iodinated molecules*** *at equal metal basis*. Synthesis and characterization of the NPs will be shown, *in vitro* studies with ER+ breast cancer cells with our NPs and controls, CT studies in solution and cell pellets, as well as results from our first *in vivo* studies with a relevant animal model, *where both active and passive tumor targeting* are explored. Our approach combines ***NP targeting receptor specificity*** with ***dual-energy CT (DE-CT) or photon-counting spectral CT (PCD-CT) for tumor and micro-tumor detection.***

275b Effect of Post-Synthesis Oxidation on Magnetic Particle Imaging (MPI) Performance

Ambar Velazquez Albino, Aniela Nozka, Andrii Melnyk, Hayden Good, Carlos Rinaldi-Ramos

Magnetic nanoparticles are of great interest for biomedical applications due to their ease of synthesis, biocompatibility, and tunable physicochemical and magnetic properties. Magnetic particle imaging (MPI) is a novel molecular imaging modality relying on the nonlinear magnetization of superparamagnetic iron oxide nanoparticle (SPION) tracers in a time-varying magnetic field, which has potential for applications in blood pool imaging, magnetic hyperthermia, magnetically triggered drug release, and tracking cell therapies. Modeling of MPI physics suggests that iron oxide nanoparticles larger than 25 nm with uniform magnetic properties would achieve optimal performance, meaning high signal intensity and resolution. The challenge of synthesizing these tracers is that as size increases, the probability of defects and multiple crystal structures forming within a particle increases, impairing magnetic properties and hindering MPI performance. Synthesis methods reported to improve magnetic properties include in-situ oxidation and post-synthesis oxidation. This study focuses on the latter using a parallel-synthesis approach to perform 8 syntheses simultaneously, accelerating data collection and enabling statistical analysis of particle properties and performance. Two control groups were synthesized without post-synthesis oxidation, meanwhile two groups underwent post-synthesis oxidation with a 1% O_2 in argon mixture at 300 °C for either 3 or 30 hours. Characterization consisted of TEM for physical diameters, and SQUID magnetometry for M-H curves to determine magnetic diameters and saturation magnetization. MPI performance was obtained from relaxation scans, both signal intensity and resolution, showing significant improvement for the post-synthesis oxidation groups. Correlations and statistical analysis suggest that the following tracer properties most strongly influence MPI performance: magnetic diameter and its distribution, difference between the physical and magnetic diameter, and physical aspect ratio.

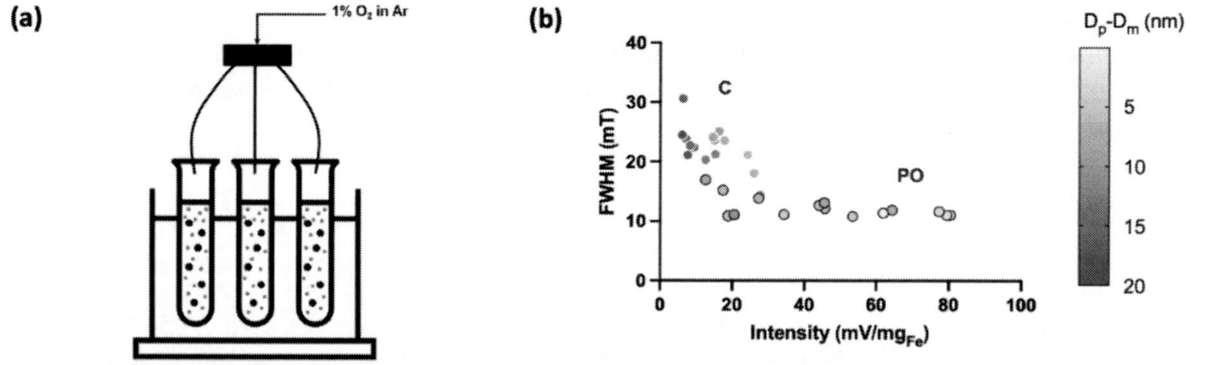

Figure 1. Synthesis of iron oxide MPI tracers and the effect of post-synthesis oxidation: (a) Synthesis setup: oxygen (red) diffuses through iron oxide nanoparticles (black) during post synthesis oxidation, (b) Effect of D_p-D_m on MPI performance (intensity and FWHM) for control (C) and post-synthesis oxidation (PO – black border) samples

275c Magnetic Nanoparticle Composite for Dual Mode T1-T2 Magnetic Resonance Imaging

Shaquib Rahman Ansari, Qianying Chen, Peter Svedlindh, Alexandra Teleki

Magnetic resonance imaging (MRI) often requires the use of contrast agents for enhanced visualization of tissue pathologies. T1 and T2 contrast agents are used to improve the T1- and T2-weighted sequences, respectively. Gadolinium-based contrast agents are used for T1 contrast enhancement, especially of fatty tissue. However, they are associated with acute nephrotoxicity. T2 contrast agents such as superparamagnetic iron oxide nanoparticles (SPIONs) are superior in imaging edema and inflammation.[1] Additionally, ultrasmall SPIONs (USPIONs) have been demonstrated to significantly enhance T1 contrast in MRI, thus preventing the nephrotoxic effects of gadolinium-based contrast agents. Development of a magnetic nanoparticle based T1–T2 dual mode MRI contrast agent can significantly improve diagnostic accuracy by providing paired anatomical images with different contrasts, and reduce drug related adverse effects. Therefore, in this study we develop SPIONs and USPIONs with enhanced contrasting ability, and combine them in situ into magnetic nanoparticle composites using flame spray pyrolysis (FSP) to obtain a dual mode T1-T2 contrast agent.

SPIONs and USPIONs were produced by FSP. The crystal size was varied between 3 to 24 nm (to yield superparamagnetic particles). The nanoparticles were coated with citrate to improve aqueous dispersibility, and characterized for their structural and magnetic properties. The T1 and T2 contrast efficiency of the nanoparticles was determined using a 9.4T Varian scanner. Finally, the SPION and USPION with highest contrasting efficiency will be combined in situ using the double-nozzle FSP technique,[2] and characterized for their structural, magnetic and MRI contrast enhancement properties.

Nanoparticles of sizes 3, 8, 16, and 24 nm were obtained by fine-tuning the FSP parameters. Citrate coating of the SPIONs and USPIONs significantly improved the aqueous dispersibility and stability. The largest SPION (24 nm) showed the highest saturation magnetization (55.6 emu g^{-1}) compared to other SPIONs, and a high T2 and T2* relaxivity of 359.4 and 409.2 $mM^{-1}s^{-1}$, respectively. Furthermore, the 24 nm SPIONs showed very low T1 relaxivity (0.164 $mM^{-1}s^{-1}$). The relaxivity values of 24 nm SPIONs indicates their effectiveness as strong T2 contrast agents.

Large SPIONs showed high T2 contrast efficiency. A double-nozzle FSP will be used to combine nanoparticles of different morphologies and contrast abilities. This study demonstrates that magnetic nanoparticle composite comprising of USPIONs and SPIONs can be used as dual mode T1-T2 contrast agent.

Acknowledgement:

This project has received funding from the European Research Council (ERC) under the European Union's Horizon 2020 research and innovation programme (grant agreement No. 101002582). The authors also acknowledge financial support from the Science for Life Laboratory.

References:

(1) Gee, M. S.; Harisinghani, M. G. MRI in Patients with Inflammatory Bowel Disease. *J. Magn. Reson. Imaging* **2011**, *33* (3), 527.

(2) Grossmann, H. K.; Grieb, T.; Meierhofer, F.; Hodapp, M. J.; Noriler, D.; Gröhn, A.; Meier, H. F.; Fritsching, U.; Wegner, K.; Mädler, L. Nanoscale Mixing during Double-Flame Spray Synthesis of Heterostructured Nanoparticles. *J. Nanoparticle Res.* **2015**, *17* (4), 1–16.

275d Exploring High-Density Lipoprotein-Based Magnetic Nanoparticles As Biomimetic MRI Contrast Agents for Cancer Diagnosis

Shiva Rahmati, Allan E. David, Xin Fan

Magnetic resonance imaging (MRI) is a powerful imaging modality widely used in the diagnosis and prognosis of various disorders including cancer. To enhance the contrast between normal and pathological sites in the body and consequently improve the sensitivity and diagnostic efficacy of MRI, exogenous contrast agents (CAs) are often administered in clinical settings. Owing to their biocompatibility, and tunable physiochemical and magnetic properties, superparamagnetic iron oxide nanoparticles (SPIONs) have received considerable attention as MRI contrast agents. Current commercially available SPION-based MRI contrast agents including Resovist and Feridex, however, suffer from low sensitivity (low contrast–enhanced properties) and poor cancer specificity. To address this issue, biomimetic MRI contrast agents based on SPIONs are of particular interest not only due to their higher tumor specificity but their good sensitivity and higher biocompatibility. Here, for the first time, reconstituted high-density lipoproteins (rHDLs) were employed as biomimetic cancer delivery agents for SPIONs. HDLs are endogenous cholesterol transporters receiving much attention in biomedical applications. Incorporating rHDLs not only enhances the stability and biocompatibility of SPIONs but also provides specific cancer delivery through targeting HDLs' receptors (scavenger receptor class B type 1 [SR-B1]) overexpressed on a variety of cancer cells including prostate cancer.

In this study, Hydrophobic SPIONs were synthesized through a thermal decomposition process, and subsequently, SPION-loaded rHDLs were prepared according to the solvent evaporation procedure. The obtained SPIONs and SPION-loaded rHDLs had the average hydrodynamic size of 15-40 nm and 100 nm, respectively, with PDI < 0.2. The synthesized nanoparticles were further characterized in terms of composition, size, polydispersity index (PDI), ζ-potential, and X-ray diffraction. The encapsulation efficiency of SPION-loaded rHDLs was determined to be 80%, and their high stability was confirmed upon incubation in 10% FBS RPMI-1640 cell culture media as a physiologically relevant microenvironment. Their superior MRI contrast-enhanced properties compared to the conventional MRI contrast agents were demonstrated using 3T MRI. Moreover, their cancer-targeting capability was evaluated by in vitro studies in which a significantly higher uptake of SPION-loaded rHDLs was shown by PC3 and LNCAP prostate cancer cells compared to the normal cells. Finally, the cytotoxicity studies demonstrated >86% viability of CHO cells and normal prostate cells after 72 h incubation with SPION-loaded rHDLs at the Fe concentration of 0–200 μg/mL validating their high biocompatibility.

All in all, this work proposed a biomimetic MRI contrast agent based on magnetic nanoparticles for cancer diagnosis. The incorporation of SPIONs into rHDLs results in a nanostructure with biomimetic properties, high stability, high biocompatibility, superior contrast-enhanced properties, and cancer targeting capability which could overcome current issues of MRI contrast agents and provide safe and efficient cancer diagnosis.

275f Tailored Synthesis of Multi-Color Emissive Carbon Dots for Efficient and Rapid Photothermal Nucleic Acid Amplification

Salar Balou, Aashish Priye

This study focuses on the development of multi-color emissive carbon dots (CDs) with systematically tailored chemical and physical properties. Through a modified temperature and acid reagent engineered hydrothermal synthesis protocol, highly stable, cost-effective, and environmentally friendly CDs were synthesized to actuate photothermal polymerase chain reaction (PCR). Optical, physical, chemical and thermal characterizations were conducted to investigate their light-to-heat conversion performance. Results showed that red-emitting CDs had significantly higher light-to-heat conversion efficiency than their blue, cyan, green, and yellow counterparts. The study further revealed that increased acid treatment during the fabrication of CDs enhances aromatization and introduces stronger electron-donating functional groups, leading to a more conjugated and oxidized carbon dot structure. This bandgap engineering strategy reduces the HOMO-LUMO band gap, destabilizing the electrons in π-π^* orbitals and increasing the frequency of electron-phonon coupling. This coupling generates thermal vibration in the carbon dot lattice, initiating temperature increase in macroscopic scale. The resulting red-emitting carbon dots yielded high heating ramp rates of 13 °C/s when illuminated with a 350 mW blue laser. Finally, these nanocarbons were employed as efficient photothermal nanomaterials to achieve ultrafast thermocycling for quantitative real-time PCR, capable of performing 30 cycles of 95 °C to 60 °C in less than 10 minutes. The study highlights the potential of carbon dots as a sustainable and cost-effective alternative to traditional PCR methods and sets the stage for the bandgap engineering and utilization of carbon dots as efficient and effective nanomaterials for future photothermal applications.

255b CFD Modeling of Droplet Formation from High Pressure Nozzles Used in Spray Drying of Pharmaceuticals

Plamen Grigorov, Robert F. Meyer

Spray drying (SD) of pharmaceuticals is a process often used to stabilize amorphous active pharmaceutical ingredients (APIs) in order to increase their solubility and bioavailability. The process typically involves atomization of a polymeric/API solution into fine droplets, followed by their immediate drying to obtain solid particles. The main properties of the spray dried intermediates (SDIs), which include shape, flowability, compressibility and stability, can be greatly affected by the SD process. A great effort needs to be invested to ensure these properties are maintained upon scale up. The main parameters that need to be maintained consistently during scale up involve droplet size distribution and overall droplet drying rate. The aim of this study was to develop a computational fluid dynamics (CFD) modeling strategy that can predict the performance of high-pressure atomization nozzles, which are typically used for large scale SD processes. These nozzles are operated at very high pressures and their performance currently can only be assessed through at-scale experiments and measurements. This work was able to successfully demonstrate that by using CFD modeling techniques it is possible to predict nozzle performance and obtain the following: liquid flow rate, spray angle, liquid velocity and droplet size distributions as a function of operating pressures. The CFD modeling was successfully demonstrated using Newtonian and non-Newtonian (shear sensitive) liquids. Results obtained from the CFD simulations showed a very good agreement with experimental data. Additional information is presented regarding viscosity measurements of shear-thinning liquids and some new considerations for successful scale-up of SD processes.

255c Comparison of Granule Properties Obtained in Two Different Twin-Screw Granulators at the Same Shear Rate

Nazareth Ceschan, Maria Balbi, Pablo Ravazzoli, Fernando Muzzio, German Drazer, Gerardo Callegari

Introduction: In recent years, the pharmaceutical industry has considered moving from the historical production method in batch to continuous manufacturing. The latter is a faster and more efficient production process and FDA has encouraged the pharmaceutical industry to implement it in order to improve product quality and reduce hold times. Among different technologies for producing oral solid dosage forms, wet granulation is a well-known technique for processing materials with poor flow properties and/or with limited tabletability properties. Wet granulation (WG) is a commonly used size enlargement process to obtain granules by starting from particulate drugs and excipients and using a liquid as ligand. Granules obtained possess improved flow properties compared to raw materials. Usually, twin-screw extruders are used for continuous wet granulation. As different extruders, with varied dimensions and process capabilities are available, direct comparison among granulators cannot be made without considering further adjustments in process conditions. Thus, in this work, two continuous twin-screw extruders (QBcon1 from Bohle and Pharma 11 from ThermoScientific) were used to obtain granules and screw speed was varied in order to obtain the same shear rate in each equipment. Granules critical characteristics were studied and compared in order to optimize the process.

Materials and methods: Ibuprofen 50 (BASF), a high-dose, hydrophobic and poorly soluble drug, was used as an active pharmaceutical ingredient (API). Lactose (Pharmatose M200, DFE Pharma), microcrystalline cellulose (Avicel PH 101, Dupont) and polyvinylpyrrolidone (PVP K30, Acros Organics) were used as granulation excipients while DI water was used as liquid ligand. Materials were blended in a 4-liters V-blender at 25 rpm and using an intensifier bar (522 rpm). Materials concentration in the blend was: Ibuprofen, 60 %; PVP, 3 %; lactose, 25.9 % and MCC, 11.1 %. Lactose:MCC ratio was 70:30, in order to reduce liquid needed for granulation. Drug content uniformity in blends was studied by UV-spectroscopy (Cary 60, Agilent Technologies) at 220 nm. Screw configuration was kept constant in each equipment and as similar as possible between them. For that, a kneading zone (KZ) of 30° for Pharma 11 and of 22.5° for QBcon1 was assembled immediately after the liquid feed port. The slight difference in straggle angle is related to each equipment specific configuration element. A second KZ of 60° for Pharma 11 and of 67.5° for QBcon1 was assembled and separate from the first KZ using conveying elements and keeping the same number of pitches and similar proportional distance in each screw. In both equipment, a chopper was put at almost the end of the screw to reduce oversized granules size. A central composite design face centered DoE with three center points was used to optimize granules obtained with the two equipment. Each independent variable was studied at three different levels. Screw speed in both equipment were selected in order to provide the same shear rate, and varied between 250 and 450 rpm for Pharma 11 and between 180 and 320 rpm for QBcon1. Liquid to solid ratio (L/S) was 0.2, 0.3 and 0.4 for both equipment, while throughput was between 0.7 and 1.1 Kg/h for Pharma 11 and between 0.7 and 1.7 Kg/h for QbCon1. Each DoE experimental point was run during 2 minutes to allow conditions stabilization and during 10 minutes to collect granules for characterization. Samples were air dried until granules moisture content was similar to the original blend. Responses studied were: a) percentage of fines (granules < 125 μm); b) oversized granules (> 2000 μm); c) API content uniformity in granules; d) granules size distribution; e) granule porosity; f) compact tensile strength (all compacted to the same relative density); h) dissolution of API form tablets; i) wettability of tablets.

Results: Granules content uniformity was very close to the values obtained for the blends. Fine fraction increased when L/S ratio decreased, while oversized fraction increased by increasing L/S ratio. Higher amounts of fines and oversized granules were obtained in the Pharma 11 twin granulator compared to the QBcon1. Porosity (calculated by determining true and envelope density of granules with -10+18 ASTM mesh size) was between 0.373 and 0.521 for granules obtained with Pharma 11 and between 0.344 and 0.451 for QBcon1. As can be seen, less porous granules were obtained by using the QBcon1. Interestingly, highest porosity values were obtained with an L/S of 0.3 in both equipment. Tensile strength, determined in tablets with 0.91 relative porosity, was around 0.14 MPa for all granules studied. Pharmaceutical properties were adequate in each selected sample studied while tablet wettability was similar.

Conclusions: DoE allowed optimizing and comparing the production of granules carrying a high-dose, hydrophobic drug, keeping the same share rate in both equipment. Results demonstrate the effect of changing equipment and can be used, to some extent, for scale up.

255f Development of DEM Model to Study the Impact of Process Parameters and Material Attributes on Twin Screw Wet Granulation Process

Tanu Mehta, Bodhisattwa Chaudhuri

Purpose:

Granulation, a widely used process in pharmaceutical, agriculture, food, and paper industry, is the technique of agglomerating particles together into semi-permanent granules (aggregates) made up of easily distinguishable original particles. In wet granulation, the binder liquid is sprayed onto the surface of a mixed granular bed, resulting in the formation of aggregates, to improve powder properties such as flowability and compactability. Despite its importance and widespread use, the process is poorly understood. In this study, a Discrete Element Method (DEM) model has been developed to evaluate the impact of different process parameters (screw speed and powder feed rate) and material attributes (powder cohesion) on continuous twin screw wet granulation process.

Methods:

In this study, we have used the dimensions of ConsiGma TSG to build the DEM model. The geometry was designed using SolidWorks (Fig. 1) and the simulations were carried out using commercial software Altair EDEM. Johnson-Kendall-Roberts (JKR) and Edinburgh-Elasto-Plastic-Adhesive (EEPA) were evaluated as the contact models for particle-particle interactions and Hertz-Mindlin model was used for particle-equipment contacts. For the process parameters, different screw speeds (100 rpm, 200rpm and 300 rpm) and feed rate (5 kg/h, 10 kg/h and 15 kg/h) were evaluated. The level of cohesion representing different L/S ratio was introduced using varying surface energy values. After completion of the simulation, the particle coordination number and particle residence time were quantified. The developed model was validated using previously published experimental data.

Results:

It was observed that the mean residence increases with decrease in screw speed and increase in cohesion. Moreover, coordination number was calculated at different conditions, and it was observed that increase in cohesion increases the coordination number indicating formation of agglomerates. Using the mean residence time distribution for the twin screw granulation, the mean residence time was observed to be approximately 2 seconds in accordance with the previously published data for low cohesion material at 500 rpm. Increase in mean residence time was observed at lower screw speeds and higher levels of cohesion.

Conclusion:

DEM simulations were performed to study twin screw wet granulation to evaluate the impact of process parameters and material attributes. According to the study screw speed, level of cohesion and feed rate significantly influence the twin screw wet granulation process. DEM modeling could be a really useful tool to develop mechanistic understanding of the overall process.

Fig 1: Geometry of Twin Screw Wet Granulator

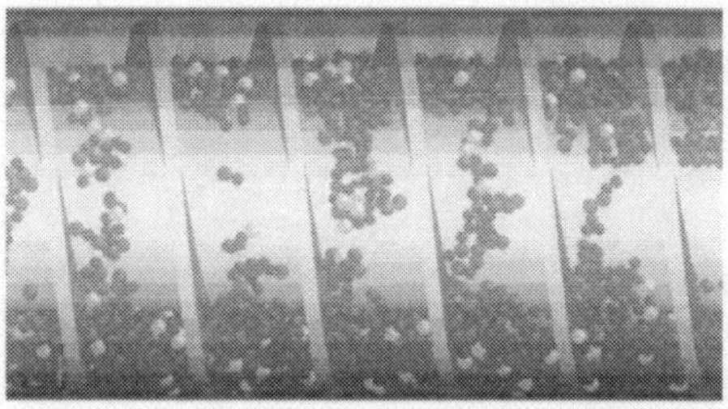

Fig 2: Twin Screw Granulator showing loose agglomerates

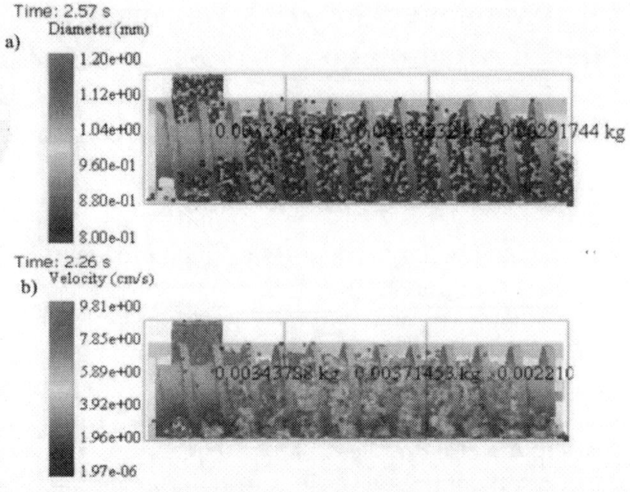

Fig 3: Twin Screw Granulation at 300 rpm with 15kg/h feed rate at high cohesion level

74b Breakage Kernels Based on 2D Analysis of High Aspect Ratio Particles

Priscilla Hill

In many industries particles are often produced in solution crystallizers where a slurry of crystals are agitated with an impeller. In the pharmaceutical industry, these crystals frequently have high aspect ratios. The high aspect ratio crystals often break more easily during agitation than low aspect ratio crystals producing fines that can significantly affect process operation, either by the fines acting as secondary nuclei or by plugging the filter immediately downstream of the crystallizer. It is noted [1] that there is a need for reliable breakage kinetics for accurate process control. The overall objective of this work is to investigate the breakage of the high aspect ratio crystals and to develop a model.

Two primary models used for high aspect ratio particle breakage represented the particles with a cuboid where the height and width of the particle are identical and are the minor axis of the particle, and the length of the particle is the major axis of the particle that is greater than the minor axis of the particle. The first model only allowed for breakage parallel to the minor axis where the child particles have the same minor axis as the parent particle [2]. The second model allows for breakage parallel to the major axis along a slip plane [3].

Laboratory scale experiments with high aspect ratio urea crystals stirred in hexane at agitation rates of 1250 and 1500 rpm provided results after 1, 5, and 15 minutes of breakage. The purpose of the agitation rates were to match the impeller tip speeds from industrial crystallizers and therefore approach the impact velocities experienced in industry. Characterization of crystals provided the necessary aspect ratio data for modeling. Previous analysis used the average aspect ratio for major axis size intervals. In general, the aspect ratio was primarily a function of the child particle major axis.

New research shows that the particle size distribution as a function of the minor axis length changes with time. The aspect ratio distribution as a function of the minor axis length also changes with time. This is not expected if breakage is only parallel to the minor axis. New 2D breakage models are presented that account for observed breakage behavior.

[1] Tyrrell, R., and Frawley, P. (2018). Single crystal fragmentation: Visualizing breakage model performance for pharmaceutical processes. *Wear*, *414*, 275-288.

[2] Capellades, G., Joshi, P. U., Dam-Johansen, K., Mealy, M. J., Christensen, T. V., and Kiil, S. (2018). Characterization of a multistage continuous MSMPR crystallization process assisted by image analysis of elongated crystals. *Crystal Growth & Design*, *18*(11), 6455-6469.

[3] Ho, R., Naderi, M., Heng, J. Y., Williams, D. R., Thielmann, F., Bouza, P., Keith, A. R., Thiele, G., and Burnett, D. J. (2012). Effect of milling on particle shape and surface energy heterogeneity of needle-shaped crystals. *Pharmaceutical research*, *29*, 2806-2816.

74c Bridging Crystallization and Mechanical Properties to Particle Size Reduction Efficiency in Jet Milling Industrial Pharmaceutical Processes

Goncalo Marcelo, André Dias, Luis Sobral, Rachid Santos, João Sequeira, Tiago Porfirio

In the pharmaceutical industry, jet milling is a preferred operation for the particle size reduction of active pharmaceutical ingredients (APIs) for inhalation products. Nevertheless, the establishment of structured relationships between the mechanical properties of API crystals and particle size reduction processes efficiency is yet scarce, often impacting scale-up, tech-transfer and process throughput [1-2]. A better understanding of this interdependency is thus essential to improve process control, reduce in-process variability and potentiate process design simplifications [3].

For that, a two-staged study at scale evaluated the impact of:

- Particle size reduction conditions on a jet milling industrial process, and link critical process parameters (CPPs) like specific energy (through pressure and nozzle orifice diameter), to critical quality attributes (CQAs) as final product PSD, morphology, and amorphous content.
- Recrystallization process conditions (i.e., water flow rate, agitation and distillation rates) in a four-variable design of experiments (DoE), linking critical material attributes (CMAs), like raw materials' PSD, crystal morphology and mechanical properties (e.g., hardness, flaws density and Young's module) to the CPPs and CQAs of the micronization stage.

With the study it is demonstrated that the energy input technique and process specific energy play an important role in reducing the number of process steps to achieve the targeted PSD, but also in reducing crystal growth after jet milling. Results show as well that different crystallizations are correlated with different Dv50 and Dv90 reductions, that vary respectively between 38 - 93% and 88 – 96 % of the initial PSD. This correlation between the physical properties of API crystals and particle reduction efficiency highlights the importance of the former as an enabler for improvements and better control over jet milling processes performance.

References:

[1] Onno de Vegt, et al., Influence of flaws and crystal properties on particle fracture in a jet mill. Powder Technology 2009, 191 (1-2), 72-77. DOI: 10.1016/j.powtec.2008.09.014.

[2] Sascha Zügner, et al., Influence of nanomechanical crystal properties on the comminution process of particulate solids in spiral jet mills. European Journal of Pharmaceutics and Biopharmaceutics 2006, 62 (2), 194-201. DOI: 10.1016/j.ejpb.2005.08.002.

[3] Mostafa Nakach, et al., Jet milling industrialization of sticky active pharmaceutical ingredient using quality-by-design approach, Pharmaceutical Development and Technology 2019, 24 (7), 849-863. DOI: 10.1080/10837450.2019.1608449

74d Breakage of Alumina Aggregates Under Shear By CFD-DEM Simulations

Eirini Goudeli, Lequan Zeng, George Franks

Controlling dispersion, aggregation and breakage of fine particles is crucial in a range of industries, including wastewater treatment, mineral processing, and ceramic powder processing, as these mechanisms can affect particle characteristics and, in turn, the process efficiency and product quality (Kushimoto et al, 2020). For instance, the size of aggregates affects the physical properties of suspensions such as viscoelasticity and particle settling rate. Aggregation of fine valuable mineral particles can enable their collection by froth flotation. Even though almost all unit operations inevitably induce shear, their effect on aggregate characteristics is not fully understood due to difficulties in monitoring the temporal evolution of aggregate behavior at short time scales at the microscopic level. This challenge can be overcome by computer simulations of particle dynamics in suspensions under different shear environments.

Here, Computational Fluid Dynamics (CFD) are coupled with Discrete Element Method (DEM) to investigate the effect of shear on alumina aggregation accounting for particle-particle (surface, contact and collision forces) and particle-fluid interactions (hydrodynamic forces). The evolving aggregate size, size distribution, and aggregate structure are quantified during concurrent aggregation and breakage induced by shear (at shear rates, G, of 300 – 3000 s^{-1}) in a Couette flow cell from perfect spherical particles all the way up to large aggregates, finding excellent agreement with experiments[1]. Snapshots of the aggregates formed by shear-induced aggregation undergoing simultaneous breakage are shown exemplarily for G = 1000 s^{-1} in Figure 1. Initially, spherical primary particles (blue spheres) collide with each other forming larger aggregates (e.g., white particles at 0.03 and 0.05 s). The particles grow further with time (red particles at 1 s) and, once steady state has been reached, the aggregation and breakage rate are both dominant resulting in internal aggregate restructuring without further particle growth. Aggregate breakage is quantified during steady state and a new equation is proposed for the breakage rate as a function of the shear rate and the aggregate size for the first time at low shear rates. Such easy-to-use breakage rate equations obtained from accurate CFD-DEM simulations can be readily used in population balance models.

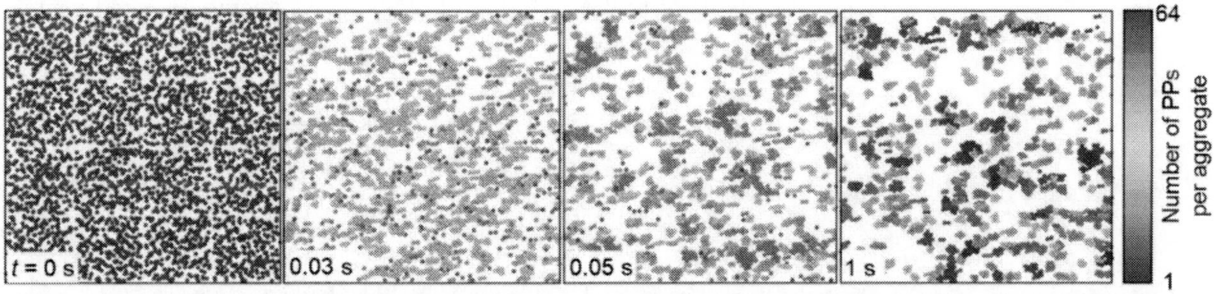

Figure 1. Snapshots of aggregates formed during shear undergoing concurrent aggregation and breakage. Only 20% of the aggregate population is shown for clarity. The snapshots are colour-coded based on the number of constituent primary particles per aggregate where blue particles correspond to small aggregates and red particles corresponding to large ones. The primary particles are 1.5 μm in diameter with 3.5 vol%.

74e Comminution of Carbon Particles in a Fluidized Bed Reactor: A Review

Sama Manzoor, Erin Bobicki

Study of comminution of carbonaceous solids during fluidized bed combustion process is essential for understanding the actual combustion rate, thermal efficiency, and particle size distribution. Comminution in a fluidized bed reactor involves a sequence of events that occur either separately or simultaneously, and with combustion process. These events are: primary fragmentation, which is attributed to stresses caused by devolatilization; secondary fragmentation, which is due to the burning up of the linkages inside the char particles; percolative fragmentation, which is a result of internal burning; and attrition, which occurs by collisions with other particles, or with the surface of the reactor. Several studies have demonstrated that due to the impact of particle comminution within the fluidized bed reactor, the carbon conversion efficiency is significantly affected. Furthermore, fine particles produced by comminution are elutriable, and they reduce the fuel residence time as well as thermal efficiency due to incomplete carbon conversion. Hence, there is a need for a complete understanding of particle comminution during practical combustion process in order to increase conversion efficiency. Considering the last review about this topic was written in 1991, the objective of this paper is to provide an updated overview of recent studies on comminution of carbon-based fuels in a fluidized bed reactor. The paper discusses different types of comminutions and their impacts on reactor efficiency, summarizes the experimental set ups used to investigate particle fragmentations, and presents the discussion on different factors of feed fuels that affect particle fragmentation. Literature experiments revealed that primary and secondary fragmentation greatly influence the fuel particle size and distribution, thus ignoring comminution phenomena will lead to errors while assessing actual fuel particle size in the reactor. It was found that fines generation and their postcombustion can cause increase in coarse char combustion rate. Carbon loss by elutriation is substantial in case of low reactive fuels only. Factors such as porosity, particle size, volatile content, and bed temperature heavily influence particle fragmentation. Fragmentation is also influenced by the initial particle size which in turn effects the particle size after de-volatilization and the size distribution throughout the reactor bed. The average size of the particles after de-volatilization and predictions of the size distributions were found to change depending on fuel type.

194a Impregnation of Catalyst Particles in a Rotating Drum with Baffles: Systematic Study Using Discrete Element Methods and Mass Transfer Theory

Maria Tomassone, Pengfei Xu, Kuang Liu

In this study we present Discrete Element Method (DEM) simulations in a rotating drum with baffles, coupled with a previously developed algorithm for fluid transfer to model dry catalyst impregnation with the goal of improving the liquid content uniformity of the resulting catalyst particles. We did a systematic study of the effect baffles to the distribution of liquid in the particle bed. We vary the number of baffles from 2 to 16, the fill level, the size and orientation of the baffles, the speed of rotation and the angle of inclination of the baffles. We also propose a mass transfer theory model to predict results theoretically. Our studies show that smaller angles give a smaller relative standard deviation (RSD) than larger angles. Smaller angles work better than straight baffles. The best case was parallel orientation with 20 degrees of inclination. Best RSD's are for the largest baffle height for all cases considered. Parallel orientation works better than opposite orientation. We also observe 2 baffles with parallel configuration with small angles have similar performance that 4 baffles straight. We also see that tilting of the baffles is equivalent to adding more baffles. If baffles are straight a larger number of baffles works better than a smaller number. Our results show that straight baffles do not generate significant convection in the axial direction. Particles behave as a rigid body in the passive zone. Wider baffles provide larger convection in the axial direction, thus smaller RSD. For straight baffles there is an optimal number of baffles. The larger the baffle height d, the smaller the Relative Standard Deviation. Baffles provide a means of accelerating the particles in the cascading zone.

The culmination of this work has been the development of simulations capable of analyzing dry catalyst impregnation in limitless parameters, geometries, and conditions for improving overall mixing, content uniformity, and product quality.

194b De-Risking and Scale-up of Mixing Baffles for Batch Corn Seed Production

Damien Brewer, Rakulan Sivanesapillai, Michael Ostendorf

Seed treatment is an efficient method of delivering formulated crop protection products where they are needed, from germination through seedling emergence and beyond. Batch processing of seed treatments is a critical unit operation within corn seed production, particularly in North America. One of the key challenges has long been chemical residue retention on the equipment itself, requiring manual cleaning by seed production site personnel. Improving the industrial hygiene of these operations is therefore a priority.

In this presentation, we describe a new commercial-scale baffle design based on pilot-scale prototyping and calibrated DEM model development to de-risk the scaleup. In-silico design testing is performed using DEM to identify the best design possible that retains mixing performance while managing baffle surface area and enhancing the projected self-cleaning performance of the mixing process. The new baffle design streamlines the cleaning process and enables additional self-cleaning capabilities while retaining strong mixing performance for corn seeds. Statistical estimators of the active ingredient distribution are taken as the measure of mixing performance, while assessments of the chemical residue "build-up" are based on cleaning time and gravimetric residue retention.

194c Design of Particles with a Prescribed Bulk Strength By Relating Particle Scale Forces to Bulk Cohesive Behavior in the Presence of Flow Aids

Kerry Johanson

Sometimes in industrial applications, we need to design a powder of limited bulk strength to prevent cohesive hang-ups (i.e. arches and ratholes). Sometimes we need to design a powder with sufficient cohesion to form good tablets. Sometimes we need to design a powder free of segregation effects. Cohesion/adhesion between particles in a mixture can play a significant role in optimal product design to prevent segregation. In general, mixtures that are cohesive tend not to have severe segregation. Finally, the design of robust agglomerates requires that a powder have the right cohesive tendency within the bulk, and how much cohesion/adhesion exists between particles plays a significant role in that agglomerate product design. In short, we want to control the cohesion in the bulk by changing something about the particle scale world and the interaction of particles within the bulk.

Unconfined yield strength is defined as the major principle stress that will cause a bulk material to fail or yield in shear when placed in an unconfined state. Bulk powder strength is defined as a bulk property, but it is caused by particle scale interactions. Thus, we can also view the bulk cohesive strength from a particle scale approach. Three or more adjacent particles can come in contact with each other and undergo a shear event. During that event, these particles move past each other in the matrix of particles that make up the bulk. As this motion proceeds the force to resist this motion is comprised of sliding friction on one of the particles and the adhesion or pull-away forces present on the other particle(s). Both forces act on the particle moving and contribute to the work required to fail or yield the bulk material in shear. Thus, by adding up the contributions of the work of the adhesion forces and the frictional sliding forces we can relate the particle scale forces to the bulk cohesive behavior. It turns out that this analysis predicts a given dependence on the particle size, depending on what causes the adhesive forces between particles. Table I summarizes the adhesion mechanisms with the particle scale strength. For the case of narrow particle size ranges, the bulk unconfined yield strength is proportional to $1/D_p$ or $1/Dp^2$ or $1/D_p{}^{1/2}$ or $1/Dp^n$ where Dp is the particle size. The exact form of the relationship depends on the cause of the adhesive forces in the bulk material.

The relationship shown in Table I is a simplification since it applies to cases where the particle sizes are uniform and real systems rarely have particles of uniform size, but are made up of particle size distributions. However, looking at a unit cube consisting of eight random neighboring particles with void space in between allows us to determine the average friction and adhesion forces to move the top assembly of particles relative to the bottom assembly of particles one particle dimension. Therefore, it is possible to use the particle size distribution to generate a series of unit cubes that represent the particles in the bulk and then compute the average work to move these particles a distance of one unit cube.

Depending on what causes the adhesion in the bulk between adjacent particles, there will be different force-work behaviors with different dependence on the particle size. Also, it is completely probable that several force-work mechanisms will be active in the failure shear motion of the particles in this unit cube. Additionally, the void space in the interior of the unit cube can be filled with smaller-sized particles which also undergo shear motion during failure. However, not all the particles in this void space may undergo shear. Some of the finer particles in the void space may be tucked into the crevasses in the adjacent particles and simply translate with the moving particles. Taking into account all of these particle interactions on a unit cube level allows us to generate a mathematical model describing the unconfined yield strength as a function of the particle size distribution and the mechanism causing adhesion between adjacent particles.

The structure of the meso-scale (i.e. the unit cube) often dictates the bulk cohesive properties. In fact, the highest bulk strength is generally caused by the condition where the interior voids are just filled with smaller cohesive particles. This should not be a surprise since it is common knowledge that cement with a well-graded aggregate gives the strongest concrete structure. One final complication occurs in the micro-scale (i.e.in the contact zone between adjacent particles). In cases of caking where crystal growth between contact points of adjacent particles is one cause of strength, very fine flow-aids can either act to decrease the adhesion force between adjacent particles or act as an aggregate binder in the solidified solid bridges formed between adjacent particles during the caking event.

The basic approach of this work is to:

1. Sieve a given material to create a series of close-sized particle sizes.
2. Measure the bulk strength of various close-size particle distributions.
3. Plot this strength data as a function of the particle diameter.
4. Fit the resulting curves to a linear combination of the strength based on each mechanism to determine the mixing coefficients and the dependence on the particle diameter.
5. Generate a mathematical model using unit cubes produced from a preselected particle size of interest (i.e. a particle size for which you wish to estimate the bulk strength). This model will include the effect of filling voids between the particles (meso-scale) with finer particles (micro-scale) and the caking time effects at particle-particle contacts with hydrophobic and hydrophilic flow aids).
6. Compare the bulk strength computed for the model with the actual measured bulk strength of the real full-size distribution material to determine the accuracy of the model.

Two particle systems are considered. The first system is a non-caking system looking at a common drug product and will predict the strength of the drug based on knowledge of the particle size distribution. The second system is a salt subject to caking, but with the addition of both a hydrophobic flow aid and a hydrophilic flow aid. The net goal of the project is to demonstrate that knowledge of the bulk strength of any real system can be computed by knowing the particle size distribution and some basic information concerning what causes adhesion between particles. It was found that the non-caking system was very well predicted by knowing the particle size distribution and the mechanistic causes of caking in the system. Good agreement was also found between the particle size distribution model and the bulk strength of the caking material. However, the hydrophobic flow aid showed a significant decrease in bulk strength versus the hydrophilic flow aid. This shows the impact that micro-scale particle contacts have on the strength prediction models.

This approach can provide formulators with a means of predicting the bulk strength of any powder using particle size distribution data and some knowledge of the causes of adhesion between particles.

Table I: Mechanisms of Bulk Yield Strength Between Particles

Strength Model for Uniform Particles	Mechanism
$fc = \frac{K_1}{Dp^2}$	Van der Waals forces
$fc = \frac{K_2 \cdot \sqrt{C}}{Dp}$	Capillary bond formation
$fc = \frac{K_3 \cdot \mu}{Dp}$	Interparticle friction
$fc = \frac{K_4}{Dp^{1/2}}$	Elastic failure of bonds
$fc = \frac{K_5}{Dp^n}$	Plastic elastic failure of bonds

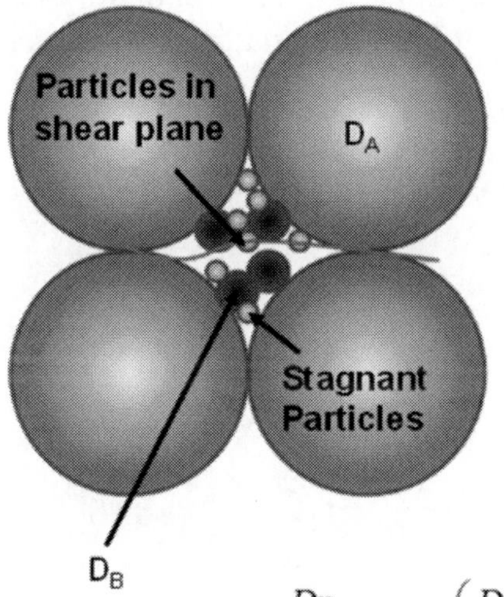

Case 3

Dp_A is the particle size of the large particles
Dp_B is the particle size of the small particles
n_{shear} is the number of particles in the shear layer between larger particles

Strength includes the effect of the forces between the larger particles and the smaller particles in the mode of shearing between the larger particles. Note some particles in the void are carried with the larger particle and do not undergo shear. These particles are not counted when computing the strength.

$$fc = \sum_A K_1 \cdot \frac{Dp_{50}}{Dp_A} + K_2 \cdot \left(\frac{Dp_{50}}{Dp_A}\right)^2 + \left[\left(K_1 \cdot \frac{Dp_{50}}{Dp_B} + K_2 \cdot \left(\frac{Dp_{50}}{Dp_B}\right)^2\right) \cdot n_{shear}\right]$$

194d Two-Component Powder Feeding Process with Twin-Screw Feeder Can be Modeled without Information of Powder Characteristics.

Yuki Kobayashi, Sanghong Kim, Takuya Nagato, Kazuhiro Uchida, Takuya Oishi, Manabu Kano

In the continuous solid dosage manufacturing process, constant supply of raw materials is important to achieve a state-of-control condition. However, precise flow rate control of an active pharmaceutical ingredient (API) is difficult to control the flow rate of powder with poor flowability. To improve the flowability, it is useful to pre-mix API and excipients with good flowability before feeding. In this study, we aim to develop a feeding process model that can be used for optimizing the appropriate mixing ratio and feeder operating conditions. The target feeding process equipped with a screw feeder feeds a two-component powder mixture of an API and an excipient. We conducted feeding experiments by changing the mixing ratio of two-component powder consisting of ethenzamide (API) and spray-dried lactose (excipient), the screw rotation speed, and the powder weight in the hopper. The experimental data were used to select an appropriate model structure and explanatory variables to predict a feed factor, which is defined as the powder feed rate per screw rotation. A feed factor is helpful to determine the appropriate operating conditions and to control the flow rate. The results showed that the gray-box model, which combines a physical model and a statistical model that expresses the parameters of the physical model, had a smaller prediction error than the other models such as partial least squared regression models, random forest models, and gaussian process regression models. The results also showed that the prediction error of the models with the screw rotation speed, the powder weight in the hopper, and the mixing ratio as explanatory variables was almost the same as that of the models with powder properties as explanatory variables in addition to these variables.

194e Improving the Properties of Pharmaceutical Powders Using Atomic Layer Coating

Shrikant Swaminathan, Kai Zheng, Miaojun Wang, Balaji Ganapathy, Neha Gupta, Fei Wang, Ankur Kadam, Pravin Narwankar

Coatings are used on Active Pharmaceutical Ingredients (API) in the pharmaceutical industry to enhance mechanical strength, enable controlled release, taste masking, reduce formulation complexity and/or to improve drug product stability. Currently, the coating technologies used on APIs have many challenges related to coating uniformity and defects, leading to variations in drug performance. This paper discusses using an ultra-thin API coating technology adapted from the semiconductor industry that can address many of the shortcomings of current coating technologies commonly used in the pharmaceutical industry. We show that this technique can achieve uniform, pin-hoe free and conformal nano-coating on multiple APIs without impacting their specific surface area and particle size distribution. In this talk, we describe how the API particles are coated and characterize the coatings. Further, we demonstrate enhancement to the powder flow properties of the drug and the ability to customize the release of the drug using vapor phase coatings. The improvement in the processability of coated API was compared against uncoated APIs in commercially relevant unit operations used in the pharmaceutical industry, such as a loss in weight screw feeder and rotary tablet press.

194f Exploiting Bacterial Adaptations for Stable Spray-Dried Enzymes- a Study of a Thermophilic Aldehyde Dehydrogenase

Wiktoria Brytan, Luis Padrela

Rationale

Spray drying is a well-established method of producing protein and peptide powders with controlled particle properties, for a span of applications including dry powder inhalables (DPIs) and continuous process implementation. One of the main drawbacks of the method is the application of heat, shear, and interfacial stresses on protein molecules, often resulting in irreversible denaturation of the product by increasing the likelihood of none-specific protein-protein interactions. Currently, stabilisation of spray dried proteins is achieved by addition of excipients. Excipients are successful in ameliorating protein stability issues, however their use can be limiting due to lung-related toxicity, product-specific interactions and their effect on final powder properties such as morphology and density [1-4]. This presentation will explore an alternative avenue for the stabilisation of proteins in spray drying by primary structure modifications. Structural mutations which enhance protein stability are widely explored in the literature however are largely ignored in their impact on solid-state stabilisation [5]. A thermophilic aldehyde dehydrogenase from *Thermus thermophilus* (ALDH*Tt*) was used as a model protein to showcase the ability of primary structure features in stabilisation of protein during spray drying. Through evolutionary adaptation, the ALDH*Tt* protein possesses a unique thermostable amino acid extension in its C-terminal [6]. The existence of this intrinsic thermostable structure in a protein, which may be excised using genetic engineering methods, cuts down on introduction of novel mutations and screening of potential thermostable mutants.

Methods

To study the effect of this structural adaptation on spray drying and solid-state stability, the C-terminus of ALDH*Tt* was excised and both proteins (with/without the C-terminal amino acid extension) were recombinantly expressed in *E.coli* BL21(DE3) cells. The proteins were then spray dried using a design of experiments (DOE) approach to statistically determine the impact of the C-terminus on the rates of enzymatic activity loss and related aggregation. The utilized DOE can be viewed in Figure 2, supplementary image. As spray drying of ALDH has not been reported in the literature, the DOE was also used to optimize the spray drying conditions of ALDH*Tt,* by monitoring the output factors of particle size, moisture content and enzyme activity. Both soluble and insoluble aggregation was assessed using Size-Exclusion High Performance Liquid Chromatography (SE-HPLC) and nephelometric turbidity, respectively. Additionally, the secondary structure conformation was explored using Circular Dichroism (CD) and Fourier Transform Spectroscopy (FTIR), both before and after spray drying. Stability of both proteins at solid-state was monitored at room temperature over a period of two months.

Results

The presence of the thermostable C-terminus extension was found to lower the activity of the native protein in aqueous solution, however it stabilised the oligomeric state of the ALDH tetramer upon solid state formation and during subsequent storage. The behaviour of the genetically engineered protein in solution has given a rationale into the evolutionary exclusion of the C-terminus from ALDH proteins in mesophilic organisms. The removal of the C-terminus extension promoted the presence of insoluble and soluble aggregation after spray drying, by allowing interaction of exposed residues with air/liquid interfaces and with other protein molecules. ALDH*Tt* with the C-terminal extension retained, on average, 24% more activity than without it during spray drying and retained up to 40% more activity during solid-state storage (see Figure 3 and 4 in supplementary image). Loss of activity from the protein lacking a C-terminus was correlated with the prevalence of high-order insoluble aggregates, and a destabilisation of the tetrameric structure by SE-HPLC and CD (see Figure 5 in supplementary image).We propose a mechanism for the protection of oligomeric proteins by the distinct C-terminal extension under shear, interfacial and thermal stresses involved in solid formation. This work is the first report outlining the effect of a structural modification on spray drying and solid-state stability of enzymes and is the first work showcasing the application of the C-terminal bacterial adaptation in the stabilisation of proteins. Although outside the scope of this work , the possibility of replicating the effect of this adaptation in other proteins using modelling and a rational mutagenesis approach may have a high impact in protein stability research. As a secondary goal of this work, the optimum process conditions for spray drying ALDH were found to be 100°C outlet drying air temperature and a feed flow rate of 1.5 ml/min using a laboratory scale spray drier Büchi B-290. Using these process conditions, dried powders of ALDH were obtained with moisture contents of 7.5%, average of 9.2 ± 2.9 μm and activity retentions of 89%. Specifically, the creation of a stable dried powder of the thermophilic aldehyde dehydrogenase may improve the breadth of this enzyme's applications in biocatalysis.

References

[1] G. Pilcer, K. Amighi, Formulation strategy and use of excipients in pulmonary drug delivery, International Journal of Pharmaceutics, 392 (2010) 1-19.https://doi.org//10.1016/j.ijpharm.2010.03.017

[2] H.R. Costantino, J.D. Andya, P.A. Nguyen, N. Dasovich, T.D. Sweeney, S.J. Shire, C.C. Hsu, Y.F. Maa, Effect of mannitol crystallization on the stability and aerosol performance of a spray-dried pharmaceutical protein, recombinant humanized anti-IgE monoclonal antibody, J Pharm Sci, 87 (1998) 1406-1411.https://doi.org/10.1021/js9800679

[3] K. Wrzosek, J. Moravčík, M. Antošová, V. Illeová, M. Polakovič, Spray drying of the mixtures of mono-, di-, and oligosaccharides, Acta Chimica Slovaca, 6 (2013) 177-181.https://doi.org/10.2478/acs-2013-0028

[4] F. Emami, A. Vatanara, E.J. Park, D.H. Na, Drying Technologies for the Stability and Bioavailability of Biopharmaceuticals, Pharmaceutics, 10 (2018) 131.https://doi.org/10.3390/pharmaceutics10030131

[5] G. Walsh, Proteins : Biochemistry and Biotechnology, John Wiley & Sons, Incorporated, Hoboken, UNITED KINGDOM.

[6] K. Hayes, M. Noor, A. Djeghader, P. Armshaw, T. Pembroke, S. Tofail, T. Soulimane, The quaternary structure of Thermus thermophilus aldehyde dehydrogenase is stabilized by an evolutionary distinct C-terminal arm extension, Sci Rep, 8 (2018) 13327.https://doi.org/10.1038/s41598-018-31724-8

Legend

ALDH*Tt* with C-terminal adaptation
ALDH*Tt* without C-terminal adaptation

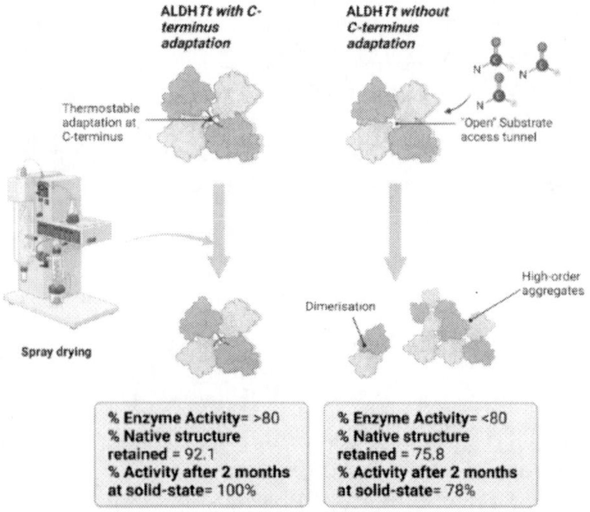

Fig.1 Graphical abstract

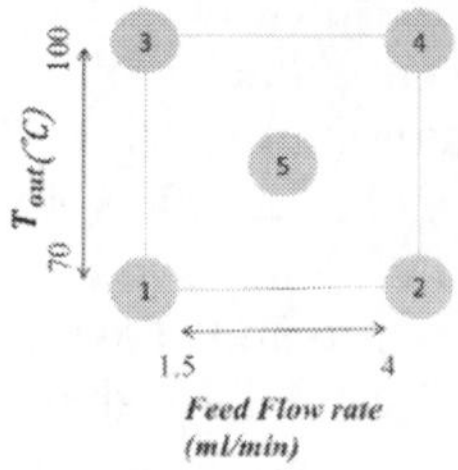

Fig.2 Design of Experiments (DOE) for the spray drying of ALDHTt

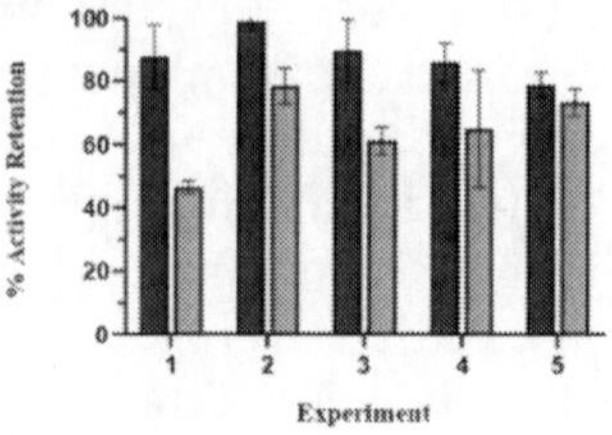

Fig.3 Enzyme activity retention after spray drying

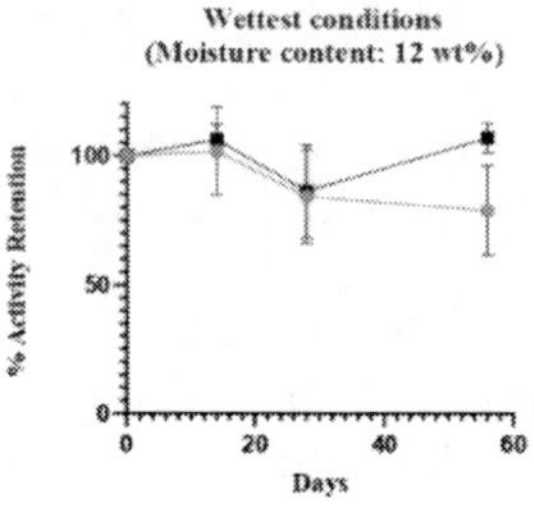

Fig.4 Stability studies

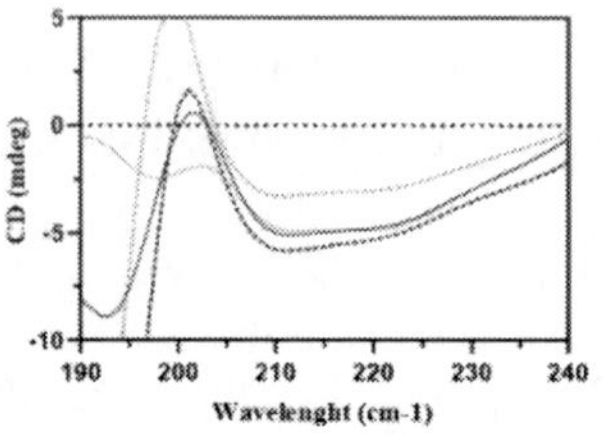

Fig.5 Circular Dichroism spectra of ALDHTt with and without adaptation

276a Segregation in Powders: An Interplay between Particle Properties and Cohesion

Aurelien Neveu, Filip Francqui, Geoffroy Lumay

Most of the powders used in industrial processes are polydisperse materials, i.e. the particle size is not unique but distributed over a range of sizes and usually described by the particle size distribution. However, when such polydisperse materials are submitted to flow or vibrations, segregation mechanisms come into play. The segregation will modify the size distribution across the material and as a consequence induce changes in the powder properties.

In lots of industries, powder batches have to meet strong requirements to guarantee good processability and the quality of the final product. The production of pharmaceutical products is a good example in regards to this, as the powder properties must comply with strict regulations to ensure the consistency of the production. The evaluation of the powder characteristics is usually done in the lab after the formulation. However, if a change in powder properties occurs when the powder goes through the different steps of the production line (feeding, mixing/blending, coating, tableting or capsule filling, ...) the processability will be affected leading to the final products being out of specification. Segregation can be one of the mechanisms at the origin of this drift in powder processing and therefore have to be properly understood to give better prediction of the actual processability of the batches. Moreover, continuous processing commonly used in production and currently under development in pharma industries is expected to highlight the problems associated with segregation.

The size distribution is one of the most important particle properties influencing segregation and probably the most studied. However, other properties such as shape and interparticle friction can lead to different segregating behaviors. The cohesive interactions lying between the particles, mainly due to the presence of capillary bridges and electrostatic interactions, play also an important role. Usually, the presence of cohesive forces will prevent an easy reorganization of the particles and thus will oppose the segregation mechanisms. However, an inhomogeneous distribution of the cohesive interactions across the material can lead to cohesion-driven segregation mechanisms. Despite the omnipresence of cohesive polydisperse powders in industrial applications, most of the studies have focused on dry non-cohesive materials.

In this study, the effect of cohesion on segregation has been investigated over a wide range of different materials. The cohesion of the powder samples is controlled via pre-conditioning at different relative humidities. The resulting change in cohesiveness is then evaluated in a quasi-static regime with a tap density analysis (GranuPack, Granutools, Belgium), and in a dynamic regime with a rotating drum method (GranuDrum, Granutools, Belgium). The segregation is investigated via two methods. Using the ability of the GranuPack instrument to measure the bulk density evolution after each tap allows us to investigate the effect of segregation on the packing dynamics. In addition, the rotating drum method used in this study allows for performing hysteresis analysis. The measurement is done by following an increasing rotating speeds sequence then following the same sequence in a decreasing order. As segregation will lead to temporal modifications of the flowability during the measurement, that will be seen as an hysteretic behavior between the two speeds sequence.

276b Segregation of Particles in Aluminum-Water Reactors Using Gas-Fluidization and Vibration

Naimah Russ, Christopher Spitler, Qiang Guo, Shahryar Jafarinejad, Christopher Boyce

This study investigates the phenomenon of particle segregation in aluminum water reactors, with a focus on the segregation of aluminum and aluminum hydroxide particles. Aluminum water reactors involve the reaction of aluminum with water to produce hydrogen gas and aluminum hydroxide. At the moment, aluminum-water reactors are a potential way to power underwater vehicles, but they are unable to do it reliably or effectively. These problems occur because aluminum particles and aluminum hydroxide are immobile and water cannot consistently reach unreacted aluminum. Although it has been demonstrated that subjecting particles to vibration and gas flow causes them to separate vertically, little emphasis has been given to the fact that vibration and gas flow produce separation in the opposite directions. In order to controllably separate unreacted particles from reaction products and cause motion in the granular aluminum particles, we use computational modeling to investigate the effects of vibration and/or gas-fluidization on segregation of aluminum and aluminum hydroxide particles. Both computational fluid dynamics - discrete element method (CFD-DEM) and multi-fluid modeling (MFM) techniques are used for simulations. Further, the effect of introducing cohesive forces to account for the effects of liquid bridging and small particle size on segregation dynamics is studied.

276c Segregation and Mixing of Passive and Responsive, Magnetically-Driven Rotating Particles

James Gilchrist, Samuel Wilson-Whitford, Jinghui Gao, David Kramer, Marietta Sisca

Gravity-driven granular flows, commonly found in nature in the form of avalanches and landslides and industrial processing, have long piqued the interest of scientists and engineers. Their actuation simplicity and response complexity have designated them as a cornerstone system of study within the soft matter and complex systems communities. Heap formation is a typical behavior of poured granular systems. With continuous addition of a granular material, such as sand, to a flat surface, as a heap forms a continuous gravity driven boundary flow can be observed in which grains are transported from an uphill to a downhill position. The heap itself grows as grains are transferred from the boundary layer to the underlying static heap. The dissipation of potential energy through frictional particle-particle and hydrodynamic particle-fluid interactions within the boundary layer shear flow leads to a fundamental slope angle, known as the angle of repose and the kinematics of the boundary layer itself have been well-documented through granular rheology of these inertial systems.

What has not previously been explored is the behavior of granular systems where shear is introduced to the system through torque applied directly to the particles themselves. In this work, a granular system of 50 micron ferromagnetic Janus particles is prepared. These responsive Janus particles are rotated through the application of magnetic torque by an external rotating magnetic field resulting in uphill granular heaping which for all purposes demonstrates quintessential granular behavior, such as boundary layer flow, but against gravity, giving rise to a negative angle of repose (downhill to uphill). When mixed with other non-functionalized particles, segregation occurs that may rise to well-known fluid instabilities. It is proposed that the behavior is governed frictional control as a direct result of the magnetic field strength. These observations elucidate a whole new way to observe granular systems and have a significant impact on the way in which granular systems could be applied.

276d Purpose-Filled-Design for Handling Systems Using a New Radial Stress Theory to Predict Segregation and Vibration Flow in Feed Systems

Kerry Johanson

At the heart of designing handling systems for bulk solids is the radial stress theory. This theory was developed in the 1960's by Andrew Jenike and Jerry Johanson as they worked at the University of Utah. The original intent of the work was to help understand the complex stresses that act in typical process vessels. However, when the radial stress theory was coupled with the concept of a radial velocity the theory was then able to predict stresses, as well as the expected velocity profiles, in these pieces of process equipment. Because of the antiquated computing capability of that day, these theories were explained and developed with the concept of limiting behavior. It became possible and easy to measure a wall friction angle and an effective angle of internal friction as well as predict a flow or no-flow criteria. Mass-flow and funnel-flow were terms coined to describe the fact that under some conditions solids may flow in only a central flow channel surrounded by stagnant zones of material. Pseudo concepts like first-in-and-last-out were used to determine process behavior and make decisions about segregation in process vessels or residence time distributions or mixing in process vessels. This approach was not very quantitative and engineers were left having to rely on pilot-scale experiments to quantify these behaviors. Inherent in the original radial stress theory were a set of simplifying assumptions that made the calculations easy to perform, but limited the usefulness of the theory to just a subset of some process vessels. Another issue about the development of the radial stress theory is that the theory was developed first and then applied to process conditions in an effort to make all observations fit the theory.

The first goal of this work is to relax these assumptions in a systematic way to extend the usefulness of the work to more extensive process conditions. The second goal of this work is to redefine what the theory provides as an output and to express this in terms of specifics of the product to be handled. Modeling and measurement go hand in hand and sometimes it is necessary to take a look back and ask the two questions:

- Given the operation protocols and geometry of the process along with the product specifications, is there a way that the theory at hand can be applied to predict something about my end product leaving my process?
- If that is possible, what must be used as a metric to describe a good product and can the theory at hand help me to predict that metric so I can make an informed judgment call on the design of my process or product?

For example, in packing I am interested in controlling the weight of material as well as the concentration of key components introduced into the package, so that I can define a metric that tells me if my product is good. I define one of those metrics by stating that the content uniformity of a product leaving the handling system must always be greater than the lower limit of allowed content uniformity and less than the upper limit of content uniformity as dictated by the customer's needs. With that definition in mind, I can use the theory at hand to compute the velocity profile in all the pieces of process equipment and then use the segregation profile to estimate what leaves the process as a function of time. I may also need to compute the residence time distribution of some parts of the process from the theory at hand and do a mixing analysis to estimate the remixing that may occur in the handling process. But, I can use the theory to compute the content uniformity leaving the process and then create a plot of content uniformity as a function of key material flow properties and process variables. Such as, the radial stress theory gives a relationship between the wall friction angle and hopper slope which provides a limiting line that determines if the material is in funnel flow or mass flow. This does not really help with the metric I have chosen to determine good or bad product, but I can use the radial stress theory coupled with the segregation profile and process description to generate a plot of expected content uniformities leaving the process as a function of the wall friction and hopper slope angle. Now we have an intelligent design tool that allows me to design the process to fit product needs. A similar thing can be done with the consistent weight metric required for the robust packing of my product. The theory, whatever it is, should be used in such a way that engineers can easily design with a purpose in mind (Purpose-Filled-Design).

All theories are based on a set of simplifications that make it easier to relate key properties to theoretical outcomes. If possible, that relationship needs to be straightforward. But the simplifying assumptions cannot be so extensive that the theory does not predict most of the real-world behavior. Therefore, it stands to reason that whatever theory is used the assumptions should be able to cover the real-world operations. The radial stress theory as originally introduced had a fairly extensive list of assumptions::

1. Limited or small accelerations of the bulk material (i.e. quasi-static flow)
2. Axisymmetric stress application and axisymmetric boundary conditions
3. A linear relationship between the stress in the process vessel and the vessel diameter
4. No external force gradients acting on the bulk material except gravity which acts in the direction of flow
5. The stress and strain rate tensors are aligned in principal directions
6. No pseudo-cohesive effects (i.e. continual deformation without volume change – flow happens without induced cohesion effects)

If flow properties are measured, then mathematical theories such as the radial stress theory can be employed to estimate the velocities in process equipment and compute residence time distributions and segregation emptying profiles for fill-then-empty processes – as well as semi-continuous processes. However, the basic assumptions of the radial stress theory are limited and many real processes do not conform to the assumptions behind the standard theory. Therefore, the solids velocity profiles computed from the radial stress theory may not be sufficient to describe segregation profiles and residence time distributions for material leaving the process.

This work examines the relationship that external gradients such as non-radial stress gradients or local gas pressure gradients in the process might have on the solids velocity profiles in process equipment. The work outlines calculations to compute velocities in bulk handling process equipment subject to these gradients and then applies those velocities to the calculation of segregation profiles. The ultimate goal and focus of the work is to relate the external force gradients to product consistency. A new radial stress theory to include non-radial stress and gas pressure gradients will be presented.

Finally, the mass flow plots created by Andrew Jenike and Jerry Johanson were used to determine velocity ratios in process equipment, feeders stress loads, and even flow factors to describe arching effects. However, they do not contain a segregation index that could be used to determine what type of mass flow will provide a certain degree of segregation prevention. The plots also do not contain a flow aid vibration index to prevent or modify funnel flow behavior. Therefore, as part of this work new mass flow plots containing segregation intensity numbers and vibration intensity numbers are generated for various non-radial solid stress gradients, local gas pressure gradients, effective internal friction angles, hopper angles, wall friction angles, and segregation patterns. These tools can assist practicing engineers in designing process equipment to more easily solve segregation problems or solve problems where the vibration of the mass is required to induce the proper flow. Depending on the direction of the excess stress gradients, gas pressure gradient, or external acceleration force, applying the radial stress theory can predict a change in mass flowability to help or hinder the flow and segregation prevention ability of the handling system.

276e Percolation of a Fine Particle in a Sheared Granular Bed

Song Gao, Julio M. Ottino, Paul B. Umbanhowar, Richard Lueptow

Granular size segregation is a phenomenon commonly encountered in various natural and industrial settings. While this problem has been extensively studied for large to small particle size ratio $R<3$ and above the geometric trapping threshold, $R>R_t=6.464$ (free sifting), segregation for intermediate R values, particularly in flowing granular beds, remains unclear. Here we investigate the percolation of a fine spherical particle in a sheared-large-particle bed with constant shear rate, γ, under gravity for $2<R<7$, which includes both the trapping and passing regimes observed in the static bed. At low γ, where the flow inertial number $I<0.01$, corresponding to quasi-static flow, a fine particle is never permanently trapped even for $R<R_t$ due to the shear-driven temporal and spatial variation of the bed structure. In this regime, the percolation velocity v_p increases with γ for $R<R_t$ but remains constant for $R>R_t$. We propose a probabilistic model that considers R, γ, particle size, gravity, and bed structure that accurately predicts v_p in this regime. However, at higher shear rates ($0.01<I<1$), the model breaks down as v_p decreases with increasing γ for all R due to the increasing granular temperature. As I approaches 1, the decrease of v_p slows, and v_p tends toward a non-zero value for all R. Additionally, a higher restitution coefficient leads to smaller v_p for all cases, consistent with results for a completely static bed of large particles. Funded by NSF Grant No. CBET-2203703.

276f Granular Segregation Forces

Yifei Duan, Paul B. Umbanhowar, Julio M. Ottino, Richard Lueptow

Segregation in granular flows occurs due to the combined effects of gravity and shear. Recent studies successfully characterized the gravity-induced segregation force in a well-mixed size-bidisperse granular mixture under uniform shear using a buoyancy-like pressure gradient term combined with a semi-empirical concentration-dependent function. However, a generalized force model is still lacking for non-uninform shear flows with locally varying shear rate and mixture concentration. Here we explore the mixture segregation force, particularly the shear-induced component, using DEM simulations of controlled-velocity flows where the shear and concentration profiles are varied. The local segregation force is quantified using a modified virtual-spring-based 'force meter' technique. Combining a previously established single intruder segregation force model that includes both pressure gradient and shear-rate gradient terms with the semi-empirical concentration dependent function produces a framework that predicts the locally varying segregation force in non-uninform shear flows of size-bidisperse particle mixtures over a range of concentrations, pressures, shear rates, and particle size ratios.

*Funded by NSF Grant no. CBET-1929265.

276g Impact of Various Feed Configurations in Continuous Mixing of Segregating Powder Mixtures

Marie-Anne Batel, Cendrine Gatumel, Henri Berthiaux

Powder mixing is employed in many industries to produce pharmaceutical tablets and capsules, food products, metal powders, cosmetics... The mixing operation has long been carried out in batch but more and more companies are moving towards continuous mixing to avoid stocks and waiting times between processing steps, as well as to ensure an easier change in production volumes and to control quality in real time. However, the homogeneity of the mixture, and therefore its conformity, cannot be guaranteed during transient states, especially during start-up.

A phenomenon of segregation between two powders of different grain sizes has been observed in the GCM500 industrial continuous mixer by Ammarcha [1]. This author found out that at mixer's start, the first particles that come out of the mixer are mainly coarse particles. Then, the composition of the mixture at the mixer's outlet gradually reaches the composition of the feed. This means that during the start-up phase, fine particles accumulate in the mixer. Indeed, the fine particles percolate to the bottom of the mixer while the coarse particles roll over the powder bed and reach the outlet quicker. It was suggested that a change in feed configuration could be pointed out as a major improvement to the process.

The objective is to study the impact of feed configuration on the mixing process for different rotational speeds and mixture compositions. The different feed configurations consist in shifting one or both feeds in space and/or in time. This intensification of the process aims at reducing the start-up time. The mixing of two powders of different grain sizes is performed in a lab-scale modular continuous mixer, based on the GCM500's geometry. The chosen powders are semolina and couscous because they are free flowing powders with a trend to segregate and are of the same chemical composition. Couscous, whose grains are larger, is colored in black. In order to measure the homogeneity of the mixture in-line, this one is poured on a conveyor belt at the outlet of the mixer and is analyzed by an image analysis system. The video is decomposed into a succession of grayscale images, which makes it possible to calculate the fraction in white (semolina) pixels for each image. The variance calculated on ten successive values of fraction in white pixels represents the intensity of segregation. Thanks to the sliding variance, it is possible to characterize the homogeneity of the mixture over time. Thus, the image analysis allows us to determine the time, denoted by t_1, for the first grains to exit the mixer and the time of the transient state, noted t_2, that refers to the time for which the fraction in white pixels becomes constant. We can then calculate $\Delta t = t_2 - t_1$ which refers to the time during which the mixture is not homogeneous. Those three time-parameters, t_1, t_2 et Δt, will allow to compare the influence of feed configuration on start-up for different operating conditions.

First, the influence of feed composition and rotational speed on t_1, t_2, and Δt are studied with a conventional feeding. The results are in agreement with Ammarcha's observations that, under certain conditions, semolina accumulates at the bottom of the mixer's inlet during startup. Then, shifted feeding is studied by shifting one or both feeders in the middle or at the outlet of the mixer. The results show that it would be interesting to also shift the feeding in time by starting one feeder later than the other. Mixing at start-up could also be improved by staging the couscous feeding along the mixer so that the couscous grains are incorporated into the mixture progressively in order to reduce the start-up time. For this, a staging device has been developed to divide the powder flow coming out of a feeder along the length of the mixer and will soon be tested.

[1] Ammarcha C., Gatumel C., Dirion J.L., Cabassud M., Berthiaux H., Continuous powder mixing of segregating mixtures under steady and unsteady state regimes: Homogeneity assessment by real-time on-line image analysis, Powder Technology, 315 (2017), pp 39-52

334b Investigation on Particle Clustering in Turbulent Channel Flow Using Large Eddy Simulations (LES)

Siddhi Arya, Partha Goswami

Particle-laden turbulent flows are prevalent in natural phenomena such as sedimentation in rivers, sandstorm, cloud formation and have applications in many industrial processes like chemical reactors, pneumatic conveying of solids, gas scrubbers. Particle response to the fluid flow depends on particle inertia, the ratio of particle size to the underlying fluid scales, etc. Turbophoresis and particle clustering causes higher local concentration of particles which leads to an increase in the probability of particle collisions and flow modification. The effect of Stokes number on particle clustering has been studied by various authors for particle-laden turbulent flow. However, the effect of other parameters such as Reynolds number and mass loading on particle clustering has not been explored in detail. In the present study, large eddy simulations are performed for vertical channel flows at different Reynolds numbers, Stokes numbers, and particle volume fractions. The preferential concentration of particles in the flow is studied by implementing the voronoi tessellation method at different wall-normal locations. Probability density functions (PDF) of cluster area for LES are found to be in good agreement with Direct Numerical Simulations. The probability for clusters to have larger area decreases for higher particle loading in the channel. PDFs of cluster area and particle concentration are used to enhance our understanding of the effect of flow parameters on cluster dynamics.

334c Continuous Feeding-Discharging Behavior of Milled Biomass in Wedge-Shaped Hoppers

Yimin Lu, Nicholas Deak, Hari Sitaraman, Yidong Xia, Jordan Klinger

The commercialization of bioenergy has suffered a long time from various material handling issues, manifested as the jamming and unstable flow of particulate biomass in handling equipment like hoppers and feeders. Fundamentally, these issues are caused by the poor flowability of biomass particles, which usually have a high aspect ratio, irregular shape, complicated surface texture, and high bulk compressibility. These material attributes set biomass apart from conventional granular materials (like sands) and make it chanllenging for existing numerical approaches to predict the flow of milled biomass. Recent studies have provided improved understanding of discharge flow behavior in wedge-shaped hoppers pre-filled with biomass. However, a more common situation in practical biorefineries is the continuous feeding and discharge, which has not yet been well studied. This study investigates the continuous feeding-discharging behavior of Loblolly pine chips in wedge-shaped hoppers by combining physical experiments and discrete-element method (DEM) simulations. A graphics processing unit (GPU) computing-enabled open-sourced DEM solver is developed with a bonded-sphere particle model to characterize the flow of elongated-shaped particles. Model parameters are calibrated from a series of laboratory-scale benchmark flow tests, and the DEM model is validated by comparing the predicted flow rate against pilot-scale hopper flow tests. Simulation studies are conducted to evaluate the effects of different feeding methods, feeding rates, and feeding heights on the effective discharge in hoppers with different flow patterns (i.e., mass and funnel flow), outlet sizes, and hopper wall inclinations. We will report the obtained simulation results and propose that the maximum feeding rate to ensure continuous discharge can be described as a function of a few hopper parameters and material attributes. This study promotes the scientific understanding of the dynamic feeding-discharging behavior of milled biomass and guides the feeding control to ensure continuous material flow and transport in biorefineries.

334d Continuous Impregnation of Catalyst Particles in a Rotating Drum Using Discrete Element Method and Machine Learning Approaches.

Maria Tomassone, Pengfei Xu, Hernan Makse, Kuang Liu

The dry impregnation of catalysts is a crucial step in the preparation of heterogeneous catalyst supports, however there has not been a lot of work done computationally on this process. In this work, discrete element method (DEM) simulations coupled with machine learning are used to study the process of dry impregnation. We have previously developed a model, which has been further validated with geometrically equivalent experiments and the results show very good agreement.

Our results show that the particle bed appears to contain two regimes. The size of the Regimes is a function of inclination angle and RPM. In Regime 1 we observe smaller inclination angles and larger mass hold up implies more forces restricting the particle movement. In regime 2 we observe larger inclination angle and/or RPM and smaller mass holdup, which implies a smaller bed height and therefore a large Regime 2. Using Machine Learning we found a function for the Relative Standard Deviation as a function of time, the angle of inclination of the vessel and the speed of rotation, and for both even and uneven flow rates: We are able to predict the full Relative Standard Deviation (RSD) vs time for intermediate values of all rotational speeds and angles within the limits we input in the LASSO algorithm. We also found a qualitative estimation of both regimes 1 and 2. In general for uneven spraying we observe a faster decay of the RSD, which should give a better product quality. Machine learning also reveals that for low RPM and low angles uneven spraying RSD is lower (better product quality), which is consistent with our previous observations of the DEM studies.

334e Quantitative Assessment of Milled Woody Biomass Attributes Impact on the Performance of Compression Screw Feeder Using Modeling and Experiment

Ahmed Hamed, Yidong Xia, Eliezer A. Reyes Molina, John E. Aston, Abhishek Paul, Mark Small, David Thompson, C. Luke Williams

Lignocellulosic biomass is an abundant and cheap source of renewable energy that can play a strategic role in shaping the clean energy portfolio to meet increasing global energy demands while satisfying the proposed measures to mitigate climate changes. However, biomass recalcitrance to decomposition to release sugar content for fermentation or upgrading adversely affects economic and efficient production of biofuels and biochemical products. This recalcitrant nature is attributed to the complex internal structure of the plant in its raw form, which gives rise to unfavorable properties, e.g., high variability, high cohesiveness, low bulk density, and characteristic flow behavior that involves complex localized elastic and plastic deformation patterns with strong dependency on the stress history. Accordingly, biomass pretreatment is mandated to change the physical properties and chemical composition of the biomass feedstocks using different mechanical, thermal, and chemical processes. Biomass poor flowability renders materials handling and feeding to the biorefinery a substantial challenge in the supply system. The unique characteristic of the lignocellulosic biomass feedstocks dictates developing dedicated design practices for handling and feeding unit operations beyond the conventional methods that are based on decades-long accumulated experience from other industries dealing with particulate materials with substantially better flow properties. Current design practices lead to operation under average feeding rates much lower than the design capacity and significantly higher operation cost due to frequent system interruptions and process upsets. Promoting an adequate understanding of the characteristic nature of the granular biomass flow behavior and the underlying transport mechanisms is still needed to reduce the overall cost and enable commercial-scale deployment of bioenergy technologies. Preprocessing and materials handling involve energy intensive processes. Thus, developing mitigating strategies requires identifying the underlying mechanisms controlling the flow behavior of this biomass feedstocks to determine critical parameters. This can aid in the optimization of preprocessing unit operations and engineering favorable flowability of biomass in feeding and handling. The present work seeks quantitative assessment of the effect of particle characteristics of milled white pine on the performance of a compression screw feeder unit that is part of the Chemical Preprocessing System (CPS) at Idaho National Laboratory using an integrated computational and experimental approach. Following quality-by-design approach, the impact of different Critical Material Attributes (CMAs) and Critical Process Parameters (CPPs) on the performance of the considered unit is studied. Three critical quality attributes (CQAs) are used as performance metrics: 1) mass flow rate, 2) shaft driving torque, and 3) specific energy consumption. Experiments and discrete element model (DEM) simulations are conducted to evaluate the impact of two CMAs – particle size and moisture content, on the chosen CQAs. The impact of a CPP, namely, shaft rotational speed (rpm) is also investigated. The results shed light on the relation between material attributes, operation parameters, and screw feeder blockage mechanisms as well as the quality of the formed plug seal.

334f Experiment-Informed Discrete-Particle Flow Models for Anatomical Fractions of Forestry Residue-Derived Granular Biomass

Yidong Xia, Zakia Tasnim, Qiushi Chen, Ahmed Hamed, Jordan Klinger, Ricardo Navar, Benjamin Davis, Oyelayo Ajayi, Tiasha Bhattacharjee

Forestry residue is an important source of biomass feedstocks in North America and often supplied to biorefineries in the form of mixture containing a variety of anatomical fractions. For example, loblolly pine residue contains a mixture of bark chips, stem woodchips, and needles. These anatomical fractions may have preferred or adverse mechanical and chemical properties for material flow and transport operations in biorefineries. Understanding the flowability of the pine residue mixture as well as the individual anatomical fractions will elucidate the optimal design of preprocessing (screening, size reduction, drying, etc.) to minimize process upsets in material handling unit operations. In this work we propose several discrete element method (DEM) particle models for the anatomical fractions of loblolly pine residue and assess their suitability for capturing the key flow characteristics measured in physical experiments. The experimental data used for calibration and validation for the DEM models include FT4 powder rheometer, tribometer, and Schultz ring shear tester. The predictability and relative computational costs of each tested DEM model will be reported. Strategies to model the herbaceous biomass and municipal solid waste (MSW) with DEM will also be discussed.

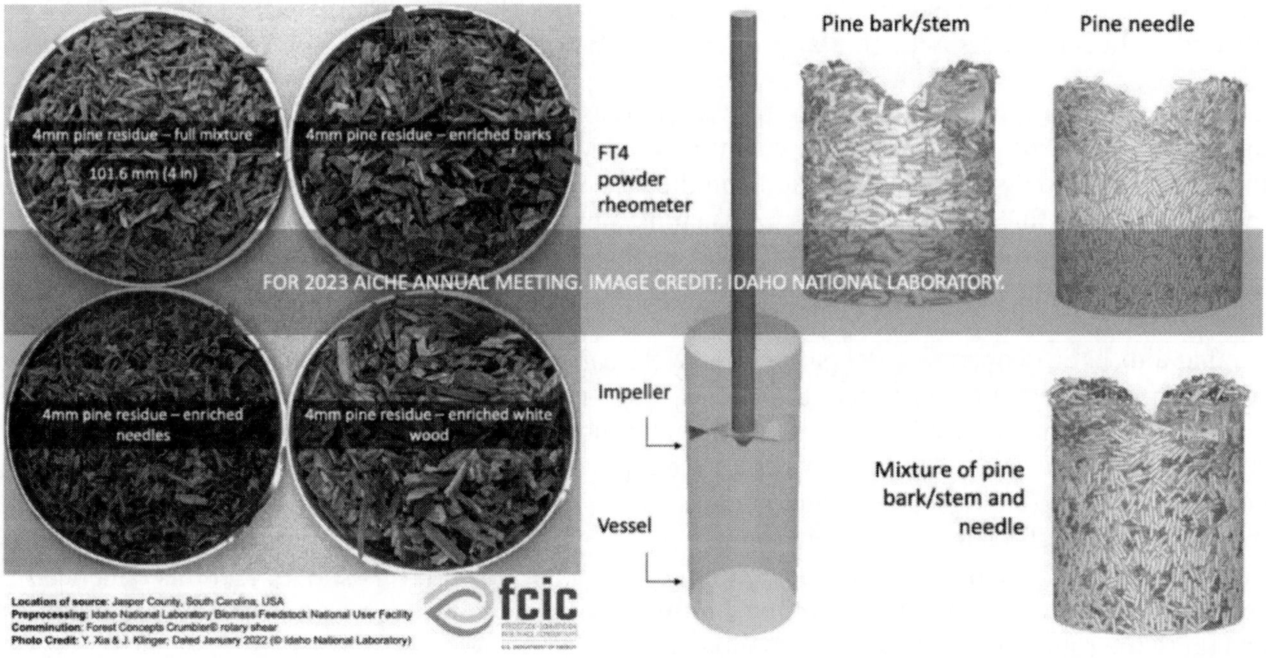

334g Evaluation of an Industrial Scale Binary Bin Blending Process Using DEM Simulations

Fatemeh Mostafaei, Benedict Benque, Thomas Forgber, Dalibor Jajcevic, Pankaj Doshi, Matthew Santangelo, Rolf Larsen, Hong-Guann Lee, Diogo Gomes Lopes, Mario Schaefer, Johannes G. Khinast

Bin blending is a unit operation commonly used in oral dosage form manufacturing to mix individual components. It is simple to implement but reaching the desired blend uniformity on an industrial scale can be challenging. One of the main challenges is to obtain the target blend uniformity in a reasonable time frame while avoiding segregation. A suitable blender and operational conditions must be selected considering the material characteristics, batch size, sensitivity to attrition, among other factors.

Several studies have investigated the mixing behavior in various types of blenders on the lab scale. Generally, it was reported that the fill level and the loading pattern have the most significant influence on blending behavior and that differences in particle size, density, or cohesivity between the materials increase the risk of segregation. The segregation patterns in different geometries and under different operational conditions have been reported in the literature. In order to achieve fast and pronounced segregation patterns, most studies have focused on lab-scale blending processes of components with very different properties (such as particle size or density). However, the risk of segregation of realistic pharmaceutical powders with similar properties in an industrial-scale process has not been addressed.

The present work studies the blending of two granular materials in a commercial scale conical-cylindrical geometry using discrete element method (DEM) simulations. The influence of geometrical modifications and operational conditions on the blending behavior and the risk of segregation were evaluated. The influence of asymmetry was studied by implementing mixing elements at the lid. The sensitivity of the process to both the fill level and repeated changes in the rotation direction was investigated.

To that aim, DEM contact model parameters were calibrated for two pharmaceutical granulates based on experimental powder characterization tests. To overcome computational limits, the particle sizes and contact parameters were scaled up while ensuring similar bulk powder behavior. For economic reasons, industrial scale blending processes are often operated at high blender fill levels. The results confirm previous findings which show increased blending times at higher fill levels. To speed up mixing, angled baffles in the lid were found helpful by improving mixing along the rotation axis, but they also increase the risk of axial segregation of the binary blend. It is shown that this enhanced segregation risk, caused by the asymmetry of the mixing elements, can be mitigated by regularly changing the rotation direction of the blender.

381a Effect of Surface Asperities on Particle-Phase Stress in Granular Flows

Alejandro Gallegos Tovar, Jiecheng Yang, Jennifer Curtis

In most industrial processes involving granular flows, such as pharmaceutical manufacturing and food processing, particles described as rounded tend not to be perfect spheres, having irregularities on the surface. In order to describe the flow behavior of these types of particles, the effect of particle surface asperities must be understood. In this study, the effect of particle surface asperities in dilute and dense-phase granular flows is investigated using discrete element method (DEM) simulations of a 3D system of particles with surface asperities in simple shear flow applying Lees-Edwards periodic boundary conditions. The particles are described by a main sphere with surface asperities depicted as hemispherical protrusions. The effect of surface asperities is assessed by progressively varying the asperity height as well as asperity abundance. Asperity height is measured as the distance from the surface of the main sphere to the top of the asperity. The asperities are distributed uniformly across the surface of the sphere. It is found that at low solid volume fractions, the asperities do not have a significant effect on the stress of the system. For larger solids concentrations, the asperities only have a significant effect on stress and flowability after they reach a critical height. For each particle simulated, the type and distribution of contacts are determined over a range of solid concentrations. The shear and normal stresses increase with asperity abundance, showing a dependence between stress and the number of surface asperities.

381b Top-Bottom Method for the Calibration of DEM Simulation Based on a Set of Complementary Measurements.

Geoffroy Lumay, Filip Francqui, Aurelien Neveu, Ben Jenkins, Kit Windows-Yule

DEM simulations are useful to predict the behaviour of powders and granular materials inside industrial processes. Conceptually, two methods can be used for the calibration of DEM simulation parameters. The first method is a bottom-up approach based on the direct measurement of the microscopic parameters at the scale of the contacts between the grains. Even if this bottom-up method makes sense conceptually, it is often practically impossible to use this approach based on complex measurement. The distribution of grain size and shape in real powders leads to strong fluctuations of the results. In addition, in many simulation models, the input parameters are not rigorously linked with these physical parameters.

We propose an alternative top-bottom method based on a set of macroscopic measurements that can be used practically without difficulties to calibrate the simulation. The principle consists in measuring the powder properties in a laboratory and afterward to simulate the same process to tune the simulation parameters following an optimisation process. These measurements are performed in well-known geometries (tubes or rotating drum) and can be easily simulated. Finally, the best set of parameters will be used to simulate a complex industrial process for example.

As for any powder characterisation task, the measurement geometry and the stress applied on the sample must be selected in accordance with the application. For a process involving high speed powder flow with a low confinement, the rotating drum methods (GranuDrum) will extract parameters that make sense for the application. On the other side, the flow through an aperture method (GranuFlow) provides a simple and quick classification of flowability. In between, the optimised GranuHeap angle of repose measurement is a good compromise. In this work, we demonstrate the potential applications of digital twins of these widely used powder characterization devices.

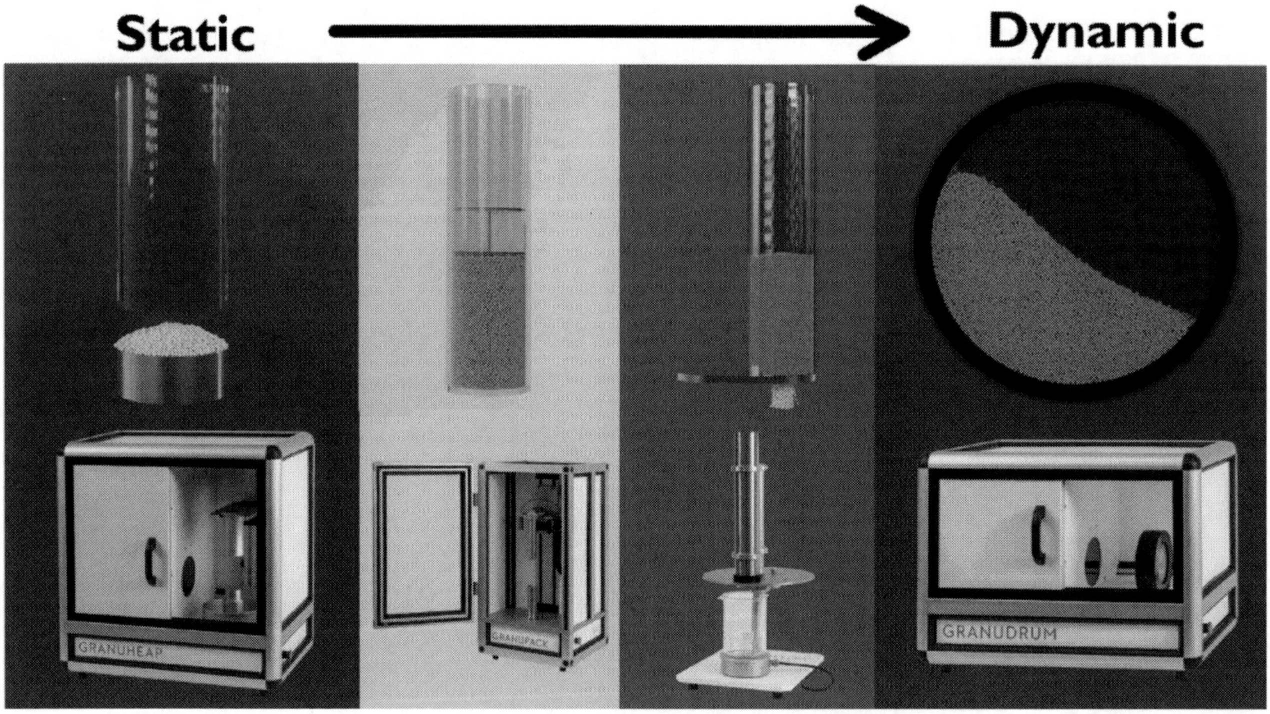

381c Continuum Modelling of Cohesive Powder Flow

Sameer Kamath, Carl R. Wassgren, Marcial Gonzalez, William Ketterhagen, Sean Garner

A continuum model implemented in the commercial finite element method (FEM) software Abaqus is used to model the powder dynamics of Avicel PH101 at low densities and low stresses, where the powder is able to flow. The modified Drucker-Prager Cap elastoplastic material model is used to model the powder. The results from shear cell testing and die compaction of Avicel PH101 are used to calibrate the plasticity-related material parameters. The elastic parameters are approximated from hydrostatic triaxial test data for Avicel PH102. A Flodex is used as a sensitive calibration experiment to resolve uncertainty in the four plasticity parameters at very low densities. The Flodex hole diameter in the simulations was set to the critical hole diameter that was experimentally determined. Then, a variance-based sensitivity analysis was carried out to determine how the amount of mass discharged from the Flodex as well as the mass flow rate is affected by varying the four parameters at low densities in the FEM simulations.

381d Discrete Element Modeling of Deformable Biomass Particle Flow in Wedge-Shaped Hopper

Qiushi Chen, Zhengshou Lai, Yidong Xia

Many natural and engineered granular materials have relatively deformable particles. Besides particle size and shape, particle deformability is another salient factor that significantly impacts the material's flow behavior. In this work, the flow of irregular-shaped deformable woody biomass particles in a wedge-shaped hopper is investigated using discrete element simulations. A bonded-sphere model is developed to simultaneously capture irregular particle shapes and particle-wise deformations (e.g., compression, deflection, and distortion). Quantitative analysis of the effects of irregular shapes and particle deformations shows that the increase in particle stiffness tends to increase initial packing porosity and decrease the flow rate in the hopper. Rigid particles tend to have clogging issues, whereas deformable particles have less chance to, indicating particle deformation reduces the critical bridging width in the hopper flow. Detailed analysis of stress fields is also conducted to provide insights into the mechanism of particle flow and clogging. Stresses and discharge rates calculated from numerical simulations are compared and show good agreement with Walker's theory and the extended Beverloo formula. Simulations with various particle shape combinations are also performed and show that the initial packing porosity decreases with an increasing percentage of fibers while the discharge rate has a complex dependency on particle shapes.

381e Modified Ergun Equation for the Pressure Drop Prediction of Packed Beds with Size-Bidisperse and Polydisperse Particles

Song Gao, Yi Fan, Jorg Theuerkauf, Kevin Kellogg

We study the gas pressure drop in gas-solid packed beds with size-bidisperse and -polydisperse spherical particles using experimental and experimentally validated coupled CFD-DEM numerical methods. For bidisperse systems, the existence of small particles can increase the overall packing density (compared to monodisperse systems), and in turn affect the length scale to characterize the flow behavior in packed beds. Consequently, the original Ergun equation using the Sauter mean diameter as the particle size input can underestimate pressure drop for all considered conditions up to 30% when compared with experimental results. Therefore, we propose a modified form of Ergun equation, where we still use the Sauter mean diameter () in the inertial term, but use (the ratio of second moment (area weighed) to first moment (number weighed) of particle diameters) in the viscous term. Because considers both particle volume (body force) and surface area (friction), it better represents more dynamic fluid-particle interactions in the inertial flow regime. In contrast, because only considers particle surface area, it better represents the friction effect, which dominates in the viscous regime. This modified model is also consistent with the original Ergun equation in the monodisperse spherical particle situation, where =. The modified equation yields substantial improvement for the pressure drop prediction. The deviation of the modified Ergun equation is generally less than 3% compared to CFD-DEM and experimental results, a significant improvement from 30% difference when is used in both inertial and viscous terms. We also validated the proposed equation for several examples of polydisperse particle mixtures. Future work needs to focus on non-spherical particle systems to further validate the modified Ergun equation proposed in this work to provide more accurate predictions for various practical applications.

381f Validation of a High-Performance Framework for Simulating the Hopper Discharge Flow of Milled Biomass

Nicholas Deak, Yimin Lu, Hari Sitaraman, Jordan Klinger, Yidong Xia

Variability of lignocellulosic biomass characteristics has been highlighted as one of the primary factors impacting feedstock bulk handling and transport, flowability, and ultimately integrated biorefinery (IBR) operational efficiency. Biomass characteristics including particle size, shape, and moisture content, which may vary considerably across common feedstocks, significantly affect flow behavior. When not accounted for, such variability can lead to unreliable IBR operation (e.g. flow arching, ratholing) and reduced production capacity, as many of the existing methods and tools developed for handling bulk solid flow were designed for relatively uniform particles with low compressibility. Numerical methods have emerged in recent years as an important tool for investigating such problems given their ability to provide detailed flow information for a wide range of problems. The discrete element method (DEM) is a widely used approach for simulating granular materials but suffers from high computational costs which often prevent its use in industry-scale simulations. In this study, we develop and apply an open-sourced GPU-accelerated DEM solver capable of performing high-fidelity large-scale simulations. The solver is used to investigate milled corn stover discharge from a wedge hopper. Glued and bonded sphere models, which represent particles as clusters of spheres that are either rigidly joined, or may bend, twist and deform, respectively, are incorporated to simulate variable particle shapes and sizes, while a liquid bridge model accounts for particle cohesion due to moisture content. Simulation parameters are obtained from a series of laboratory-scale experimental flow tests. The DEM solver is extensively validated by performing a series of axial compression tests followed by simulations of a pilot-scale wedge hopper pre-filled with corn stover. The discharge rate and arching distance are measured and compared against experimental results and are found to be in good agreement after careful parameter calibration. The validated model is then used to investigate the impact of important particle characteristics (e.g. particle size, shape, and moisture content) on bulk flow behavior. Finally, to assess the solver's readiness to simulate industry-scale flow, the GPU speedup is compared against traditional CPU-based parallelism and it is observed that the performance enhancement when using GPU acceleration is significant, allowing for commercial-scale simulations of feedstock handling.

404b The Effect of Cohesion on Particle Flow and Heat Transfer in a Rotary Drum

Carlin Leung, Elaheh Ardalani, Marcella Alves, William Borghard, Benjamin Glasser, Alberto Cuitiño

There is a wide range of granular materials used in daily items like building materials, chemicals, pharmaceuticals, and food. Sand, sugar, corn, wheat, salt, peanuts, flour, cereal, fertilizers, wood chips, catalysts, and pills are just a few examples. Rotary drums are the most common equipment used for the thermal treatment of solids and are used for both free-flowing and cohesive materials. A better understanding of heat transfer in rotary drums can improve the quality of products as well as save energy and materials. For good product quality, it is often necessary to raise the temperature of the particles uniformly. In general, materials may be cohesive due to their inherent nature or may be cohesive due to the presence of water or other binding agents. Studies have been done on the mixing of cohesive materials in rotary drums. However, to the best of the authors' knowledge there are no studies concentrating on heat transfer of cohesive materials in the rotary drums.

Understanding the relationship between cohesive particle properties and rotary drum operating conditions on the heating time is important for predicting processing time in real-world applications. In this work simulations using the discrete element method (DEM) and the JKR cohesion contact model were carried out as a means of better understanding the effect of cohesion on heat transfer. The model takes particle-to-particle and particle-to-wall interactions into account. It is based on the pull-off force due to the surface energy of particles and the van-der-Waal force curve regularization. The operating conditions were altered by adjusting the following variables: particle fill level and speed of rotation. Furthermore, the effect of material parameters was investigated by varying the surface energy, size, and thermal conductivity of the particles.

The angle of repose can indicate the cohesiveness of the granular material, referring to the Carr classification of flowability. In order to calibrate the effective surface energy of the different DEM models running in the rotary drum, an angle of repose (AOR) funnel experiment was simulated. The results demonstrate the effect of particle cohesion on the rate of heat transfer as a function of fill level, thermal conductivity and coefficient of friction. Generally, increasing particle-to-particle cohesion increases the thermal time as particles form low density clumps which leads to lower rates of heat transfer. Generally increasing particle-to-drum cohesion decreases the thermal time as particles stick to the drum wall; however, this may be undesirable due to the non-uniformity of the heating. At higher fill, the effect of wall cohesion is less important since there is a lower fraction of particles at the wall but at low fill, wall cohesion dominates the heat transfer effects.

404c Powder Tribocharging: Measurement, DEM Simulation with Patchy Particles and Process Optimization

Geoffroy Lumay, Nicolas Preud'homme, Eric Opsomer

When two objects are rubbed against each other, they charge because of the triboelectric effect. It occurs as well between insulators as between metals whether they are made of the same material or not. This effect is predominant in powders which show significant charging due to the large rate of collisions between its constituents.

In this presentation, we first show how to measure the ability of a powder to charge electrostatically during a flow in contact with a selected material. The GranuCharge measurement method is used, and some cases studies will be shown to illustrate the measurement.

Afterward, we consider a theoretical model based on the existence of acceptor and donor sites (also called patches) of charges at the surface of the particles. This model has been implemented in a customized DEM code. The patches at the surface of spherical particles are defined using a Voronoi tessellation. These patches could be donors or acceptors of charges with a donor probability and a uniform distribution to assign the patches' nature. To test the model, we placed electrically neutral and equal in size particles in a rotating drum and measured the total transferred charge over time while the drum is rotating. The total transferred charge increases rapidly to saturate. Therefore, the patches are charging inducing the formation of agglomerates as observed experimentally. We reproduced also numerically the charging of binary mixtures with small and large grains charging with an opposite charge. One more time, this result is coherent with experimental observations.

Finally, GranuCharge measurement method and the DEM code are combined to better understand powder processing methods influenced by electrostatic charges. In particular, the electrostatic interaction is used to obtain controlled agglomeration processes.

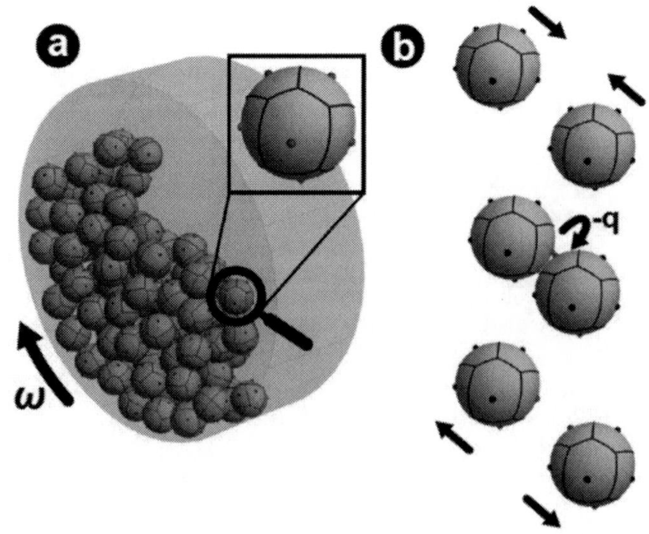

Implementation of a Pneumatic L-Valve Pulsation Scheme to Achieve Low Particle Flow Rates

Paper ID#: 668651

Michael Hayes[1], Philipp Schimmels[1], Kelvin Randhir[2], James F. Klausner[1,2], Andre Benard[1] and Joerg Petrasch[2]

(1) Mechanical Engineering, Michigan State University, East Lansing, MI, (2) RedoxBlox Inc., San Diego, CA

Abstract

Pneumatic L-valves are useful for transporting solid particles and powders without the abrasive effects of mechanical conveyors. While most L-valve research has focused on large particle flow rates and continuous aeration, achieving low solid flow rates using traditional techniques has proven challenging. This issue is especially acute when dealing with relatively large (classified as Geldart D, with d_p > 1mm) or irregularly shaped particles. This work explores an approach wherein a gas pulsation scheme achieves particle flow rates below those produced by continuous aeration. Trials are carried out on an experimental setup featuring a vertical bed of Geldart D particles (d_p = 3.66 mm), horizontal pneumatic L-valve, and real-time weighing apparatus. Gas pulsation duty cycle and frequency are identified as key parameters that characterize solid flow behavior. A linear relationship is observed between pulsation duty cycle and the average solid flow rate, while pulsation frequency is linked to the "smoothness" of the solid flow (minimization of deviations from the average flow rate). The real-time particle weighing system is further leveraged to implement a simple proportional controller to improve performance. The controller is demonstrated with varying gains to illustrate its effect on solid flow behavior. The pulsating L-valve scheme has been successfully deployed in bench-scale thermochemical energy storage systems and has the potential for broader applicability.

Introduction

The conveyance of solid particulates is a common phenomenon in industrial applications. While mechanical devices exist, pneumatic means of controlling solid flow are popular due to their lack of abrasion, applicability to high temperature environments, and general ease of use [1]. The L-valve is among the most popular of these non-mechanical approaches; its name comes from the telltale "L" shaped junction through which particles are passed. An L-valve system is depicted in **Figure 1a**. Gravity carries particles downward through the inlet of the valve, after which horizontal aeration forces particles along the length of the valve and ejects them through the outlet. Due to continuous aeration, this process facilitates a relatively constant flow of particles through the valve. Knowlton and Hirsan [2] established the first theoretical basis for L-valve operation, defining general behavior and geometric effects. A number of other authors have expanded upon this work, offering additional experimental data and flow correlations [3–7]. Studies have been performed on a variety of materials and particle sizes, ranging through the Geldart particle classifications of A [8], B [9], and D [10].

A common and well-investigated application of L-valves is in fluidized bed systems. Numerous studies have considered this use case [11,12], including high-temperature applications [13,14]. L-valves operating in fluidized bed systems exist across a variety of different industrial processes, from conveying spent coffee grounds [15] to facilitating chemical looping combustion [16]. Thermochemical energy storage is another field in which fluidized beds have some applicability [17]. Another popular reactor design is the moving bed configuration, in which particles are conveyed downwards without the active circulation of a fluidized bed. Recent moving bed thermochemical reactors rely on either mechanical means [18,19] or gravity [20,21] to maintain particle progression. While these are viable options, they introduce challenges to system robustness. In the case of mechanical, augur-like control mechanisms, particle abrasion and fragility at high temperatures impose limits on operation; for gravity-fed designs, there may be no way to control the particle flow rate beyond the predetermined feed angle.

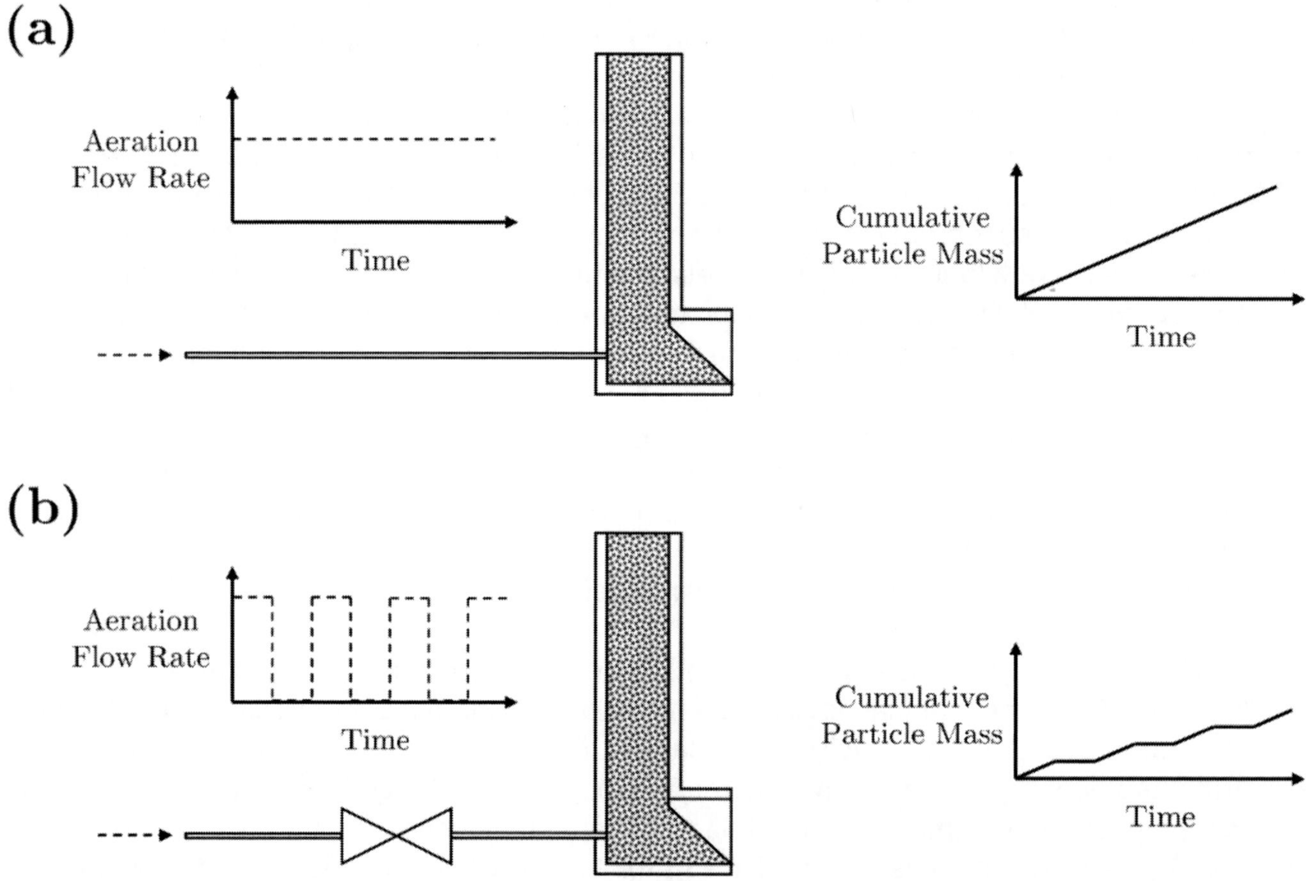

Figure 1: L-valve systems (a) without gas pulsation and (b) with gas pulsation

Although a traditional L-valve could remedy these issues, the minimum mass flow rate possible is likely too high for these applications. Particles are conveyed by the drag force of the moving gas, a drag force which must overcome the static friction of the particles themselves. Once this is achieved the particles move at some threshold flow rate, halting again once the drag force fails to overcome kinetic friction forces. As a result, a wide array of particle flow rates is unachievable using constant aeration.

Randhir et al. [22], on the other hand, proposed the use of an L-valve for conveyance of a moving bed thermochemical reactor of magnesium-manganese oxide particles. Specifically, a pulsing regime was utilized to achieve particle flow rates below that of the minimum rate for continuous aeration. Such a pulsation concept was previously considered by Vogel et. al [23] for low-flow combustion applications but has not been rigorously investigated since. This approach is depicted in **Figure 1b**. The addition of an on/off solenoid or gas flow controller to the standard L-valve system allows for the aeration gas to easily be pulsed in a square wave profile. The rate at which solid particles are ejected, then, is expected to be a fraction of the solid flow rate achieved with continuous aeration.

The general performance and characteristics of this pulsed flow are determined by two key pulsation parameters: duty cycle and frequency. As Vogel et. al [23] originally illustrated, varying duty cycle impacts average solid flow rate. Their results suggest that frequency has only small effects on solid flow, but the authors' experimental apparatus could only measure time-averaged rather than real-time particle mass data. This work seeks to expand on the results of Vogel et. al [23], including expanding pulsation characterization to consider smoothness of flow and implementing a proportional solid flow controller. The experimental system described herein allows for all of the benefits of L-valve conveyance – applicability at high temperatures, low particle abrasion, and a high degree of control – at mass flow rates that are currently only achievable through mechanical means.

Methods

Experimental Setup

The experimental setup utilized in this work is shown in **Figure 2**. An L-valve with outlet diameter of 25.66 mm and outlet length of 20 mm is mounted at the bottom of a larger feed pipe. Air is introduced from the back of the valve via a protruding nozzle. A simple on/off gas flow controller lies upstream from the nozzle. The mass of particles ejected by the L-valve is measured by a collection bin suspended from an Omega LC703-10 load cell. The bin is comprised of a steel funnel to capture particles and an electric solenoid-plug attachment on the bottom. During normal operation the solenoid remains disengaged, holding the plug in place to allow particles to collect in the funnel. When approaching the load cell's maximum weight, the solenoid is activated, lifting the plug and allowing particles to drain out of the funnel. The entire system is enclosed in a larger pressure vessel, which serves to accumulate particles discarded by the collection bin. The pressure vessel is one component of a larger thermochemical energy storage system previously described in Randhir et al. [22]. Gas flow controller actuation and recording of load cell readings are handled by an NI-USB 6218 data acquisition system.

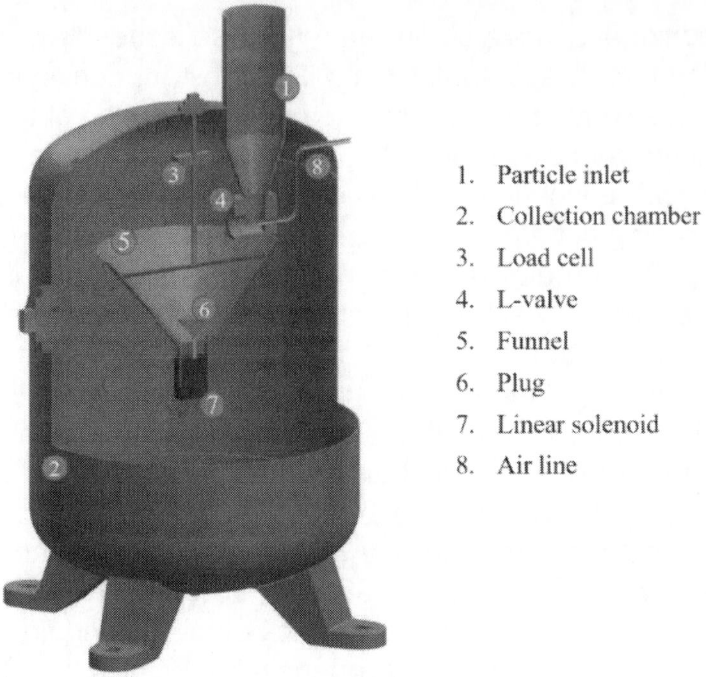

Figure 2: The L-valve experimental setup

Evaluation of Gas Pulsation Characteristics

Characterization of the pulsation parameters is carried out using roughly spherical 3.66 mm particles comprised of a magnesium-manganese oxide, the same considered in Randhir et al's [22] previous work. Particles are poured into the inlet tube. Frequencies of interest are selected; for this investigation, these values are 0.25, 0.5, 0.75, and 1 Hz. Two experimental campaigns are carried out: Session #1, which features high inlet pressure and a continuous aeration mass flow rate (duty cycle of 100%) of 148.59 g/s, and Session #2, with lower inlet pressure and a continuous aeration mass flow rate of 62.64 g/s. During Session #1, the duty cycle of pulsation is set to 0% and gradually increased to 20% in order to characterize behavior at lower values. Session #2 features a wider range between 0% and 100% duty cycle. At each increment, the system operates until 500 g of particle accumulation is recorded. This arbitrary benchmark provides the L-valve sufficient time to progress through many cycles.

Two performance metrics are calculated to evaluate the impact of duty cycle and frequency on solid flow. The first is a simple determination of the average solid mass flow rate:

$$\dot{m} = \frac{\Delta m_{\text{interval}}}{t_{\text{interval}}} \tag{1}$$

Where $\dot{m}$ is the average solid flow rate, $\Delta m_{\text{interval}}$ is the cumulative mass of particles accumulated over the measurement interval, and t_{interval} is the time of the measurement interval. The second metric considered is an expression quantifying the relative smoothness of the solid flow:

$$\gamma = \frac{\sigma(m_{\text{set}}(t) - m_{\text{measured}}(t))}{\dot{m}} \tag{2}$$

Here γ is a smoothness coefficient, t denotes quantities that are a function of time, and σ signifies the standard deviation of the mass error during operation. This error is quantified here as the difference between the expected mass measurement at a given moment for a constant solid flow rate (m_{set}) and the actual mass measured at that same moment (m_{measured}).

Deployment of Proportional Controller

Determination of a duty cycle-particle flow relationship provides a means for producing a desired particle mass flow rate simply by adjusting the duty cycle input to the aeration device. While this relationship can be used as a feedforward controller to dictate particle flow, the implementation of a feedback controller offers a higher degree of system robustness. Sinclair [24] previously applied control theory to a non-pulsating L-valve system, utilizing correlations relating aeration rate to particle flow rate. Rather than directly adjust aeration rate, this work seeks to manipulate pulsation duty cycle to control particle flow. A diagram of such a control scheme is depicted in **Figure 3**. Based on the average particle flow rate desired for the system, one can estimate what the setpoint ($m_{\text{set}}(t)$) should be over time. The feedforward relationship developed between duty cycle and solid flow (A) is applied to the derivative with respect to time of the setpoint (i.e., the desired particle flow rate). A proportional gain (K) is applied to the error between the measured mass ($m_{measured}(t)$) and the setpoint. The combination of feedback and feedforward components dictates the pulsation duty cycle delivered to the L-valve.

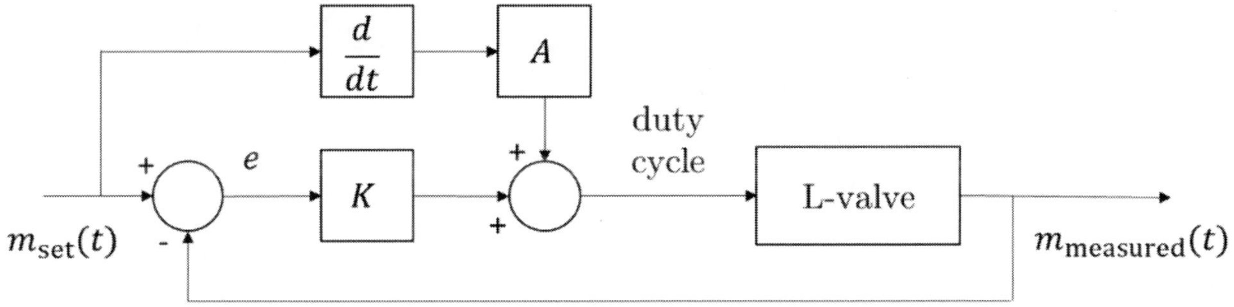

Figure 3: The control diagram for the feedback L-valve controller

Implementation of this strategy is possible due to the design of the previously described experimental setup. Real-time weighing of particle accumulation allows the system to respond to irregularities from the expected feedforward behavior. The 3.66 mm magnesium-manganese oxide particles considered in this work, with variations in diameter and sphericity, are ideal candidates to demonstrate the effectiveness of the feedback controller. Furthermore, the draining mechanism allows for a high-accuracy load cell and long operation, advantages not possible when weighing particles in a larger collection bin. Experiments are carried out to evaluate the proposed control scheme. An operating frequency of 1 Hz is chosen, as well as a desired solid flow rate of 8 g/s. The proportionality parameter K is adjusted between values of 0.01, 0.05, 0.1, and 0.5 to determine its effect on solid flow.

Results

The data collected to quantify the effects of duty cycle and frequency on particle flow are shown in **Figures 4** and **5**. Both experimental sessions are represented in the figures. **Figure 4** showcases a strongly linear relationship between duty cycle and fraction of solid flow from continuous aeration; the coefficient of determination between the recorded data and a 1:1 correlation between the two quantities is 0.985. Variability still does exist within the data set, as evidenced by points from Session #1 generally being found above the 1:1 line and points from Session #2 below it. This can be attributed to irregularities among the solid particles and other minor inconsistencies in system operation between the two periods.

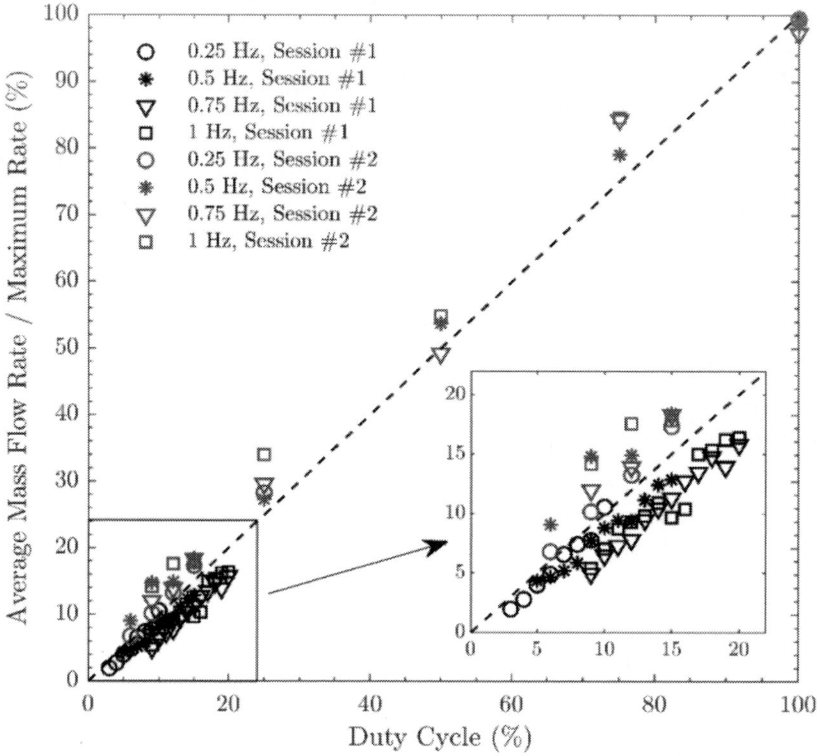

Figure 4: Results depicting the average mass flow rate produced by a fixed duty cycle at various pulsation frequencies. The dashed line represents a 1:1 relationship between duty cycle and percentage of maximum mass flow rate

An additional phenomenon of note in **Figure 4** is the minimum duty cycle available at each frequency. The lowest frequency, 0.25 Hz, shows the lowest demonstratable duty cycle in Session #1 at around 3%. It is followed by 0.5 Hz near 5%, and eventually 0.75 and 1 Hz closer to 9%. At duty cycles beyond these values all frequencies showcase consistent operation. The limitation is a result of hardware constraints, in particular the gas flow controller. If the controller is unable to switch on and off quickly enough, it will be unable to deliver aeration at low duty cycles. Low frequencies alleviate this issue by increasing the time allowed for aeration. To successfully achieve pulsation at high frequencies with lower duty cycles than are shown here, a more responsive gas flow controller would be required.

Figure 5 provides insight into how pulsation frequency impacts the smoothness of flow. The vertical axis in the figure measures the previously introduced smoothness coefficient. A larger value of this normalized error signifies a choppier, less smooth flow. This choppiness manifests as a more jagged mass accumulation profile, featuring more steps and dramatic increases in cumulative mass and less faithful emulation of constant flow. As **Figure 5** illustrates, flow smoothness is strongly affected by pulsation frequency and average solid flow rate. Across both sets of experiments, higher frequencies reduce the observed mass error; the faster the L-valve pulses the more closely the solid flow profile emulates continuous flow. Higher average mass flow rates replicate continuous flow in a similar way.

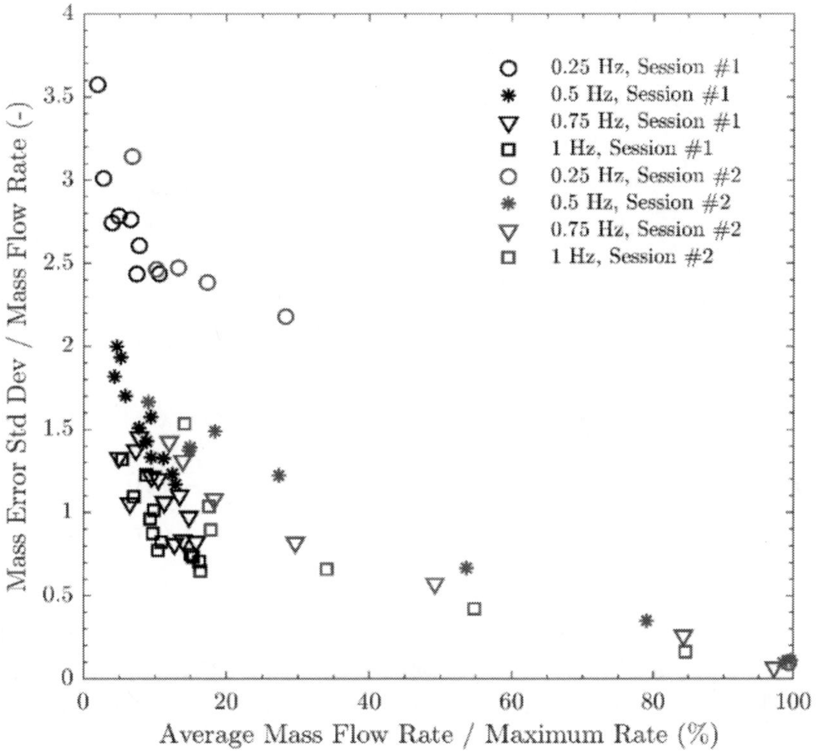

Figure 5: The standard deviation among mass errors at different average mass flow rates and frequencies

Performance of the feedback controller is shown in **Figure 6**. The impact of different tuning constants is readily apparent in **Figure 6a**, where values of K = 0.05 and K = 0.1 produce solid accumulation profiles that best track the desired output. With K = 0.01, the system is not responsive enough to discrepancies between the actual and desired mass value, leading to a large and persistent offset. A high value of K = 0.5, on the other hand, produces a choppy flow profile characterized by overcompensation. **Figure 6b** further depicts the errors from each trial. Of particular note is the difference between K = 0.05 and K = 0.1. While seemingly imperceptible in Figure 6a, the overall error for $K = 0.1$ is smaller and suggests improved performance over its competitor. The results suggest that additional tuning may help to decrease errors further.

The control approach illustrated here is just one of many strategies that may be utilized by the proposed real-time measuring L-valve system. Methods such as varying the aeration rate rather than duty cycle can likely be used to produce similar results, although this method may struggle to attain low solid flow rates compared to a duty cycle-varying approach. Furthermore, a simple proportionality gain may not provide sufficient performance. In the reactor system designed by Randhir et al [22], a more complicated control approach was implemented, featuring a pulse-width modulated input signal and gains with both proportional and exponential components. Discussion of such an approach is beyond the scope of this work, but future work may quantify the effects of more complex control strategies.

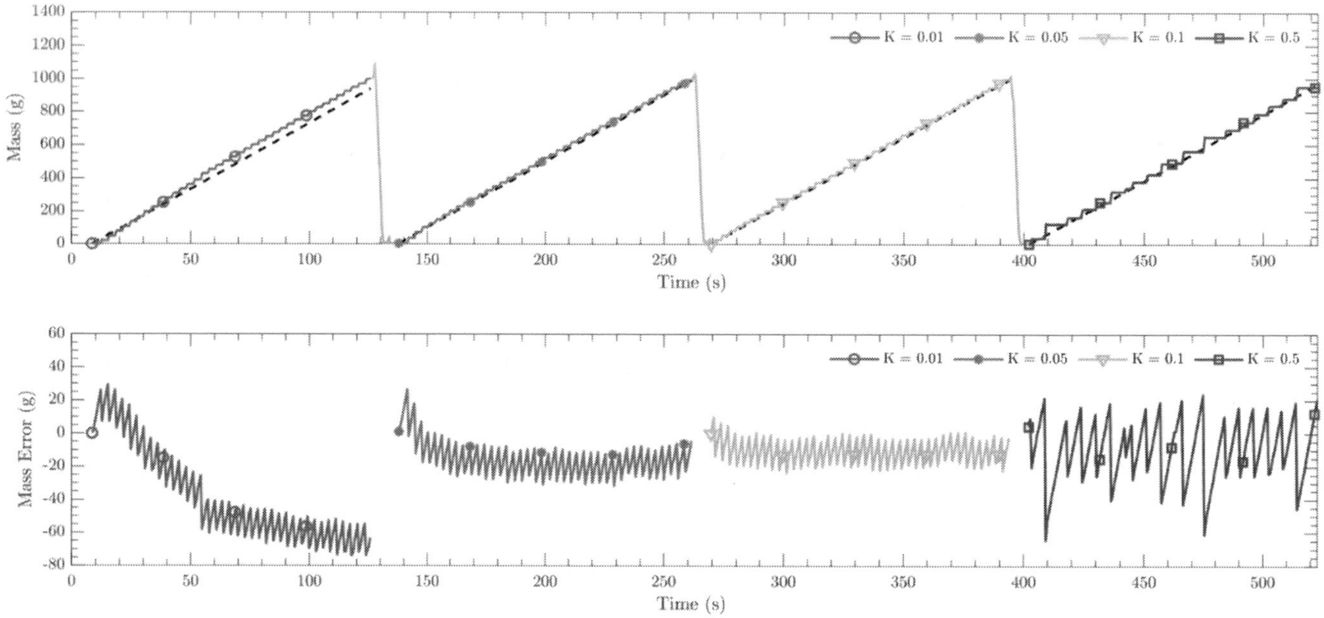

Figure 6: Proportionality constant tuning for feedback control at a desired solid flow rate of 8 g/s, with (a) real-time measurement of mass accumulation and (b) real-time measurement of mass error from the desired profile. The dashed lines in (a) represent the desired 8 g/s profile

Conclusions

This work has successfully demonstrated and characterized a pulsating L-valve system with real-time mass measurement. Duty cycle and frequency of pulsation play important roles in determining system behavior, with a suggested 1:1 duty cycle to fraction of continuous aeration solid flow relationship. Frequency impacts flow smoothness, with higher frequencies more closely emulating continuous flow. Implementation of a simple proportional feedback control allows the system to attain an even higher level of performance.

The system has already seen successful implementation in thermochemical energy storage reactors, but the approach is applicable to a nearly limitless number of use cases. Vogel et. al [23] originally considered pulsation for combustion reactors, but the described system may be applied to almost any low-flow particulate process, including those in the pharmaceutical and food industries. Future work may consider alternative control strategies and extending the results of this work to a wider range of particle sizes, materials, and use cases.

References

1 Arena, U., Langeli, C.B. and Cammarota, A. (1998) L-Valve Behaviour with Solids of Different Size and Density. *Powder Technology*, Elsevier, **98**, 231–240. https://doi.org/10.1016/S0032-5910(98)00058-8.

2 Knowlton, T.M. and Hirsan, I. (1978) L-VALVES CHARACTERIZED FOR SOLIDS FLOW. *Hydrocarbon Processing*, **57**.

3 Geldart, D. and Jones, P. (1991) The Behaviour of L-Valves with Granular Powders. *Powder Technology*, Elsevier, **67**, 163–174. https://doi.org/10.1016/0032-5910(91)80153-A.

4 Daous, M.A. and Al-Zahrani, A.A. (1998) Modeling Solids and Gas Flow through an L-Valve. *Powder Technology*, **99**. https://doi.org/10.1016/S0032-5910(98)00062-X.

5 Bhowmick, S., Baveja, N.A., Varshney, G., Thomas, D., Shringi, C.P., Shenoy, K.T., Rao, H. and Ghosh, S.K. (2013) Experimental Study on Solid Discharge Rate from a Nonmechanical Solid Feeding System. *Industrial and Engineering Chemistry Research*, **52**. https://doi.org/10.1021/ie4015028.

6 Yang, W.C. and Knowlton, T.M. (1993) L-Valve Equations. *Powder Technology*, **77**. https://doi.org/10.1016/0032-5910(93)85006-U.

7 He, P. and Fan, L.S. (2020) A General Methodology and the Correlation for the Prediction of the Solids Flow Rates through the L-Valves. *Powder Technology*, **360**. https://doi.org/10.1016/j.powtec.2019.09.090.

8 Lim, J.H., Lee, D.H., Chae, H.J. and Jeong, S.Y. (2013) Pressure Change and Control of the Solid Circulation Rate of Geldart A Particles in a Small Diameter L-Valve. *Powder Technology*, **243**. https://doi.org/10.1016/j.powtec.2013.03.046.

9 Monazam, E.R., Breault, R.W., Shadle, L.J. and Weber, J.M. (2018) Analysis of Solids Flow Rate through Nonmechanical L-Valve in an Industrial-Scale Circulating Fluidized Bed Using Group B Particles. *Industrial and Engineering Chemistry Research*, **57**. https://doi.org/10.1021/acs.iecr.8b01051.

10 Smolders, K. and Baeyens, J. (1995) The Operation of L-Valves to Control Standpipe Flow. *Advanced Powder Technology*, **6**. https://doi.org/10.1016/S0921-8831(08)60525-7.

11 Yazdanpanah, M.M., Forret, A., Gauthier, T. and Delebarre, A. (2012) An Experimental Investigation of L-Valve Operation in an Interconnected Circulating Fluidized Bed System. *Powder Technology*, **221**. https://doi.org/10.1016/j.powtec.2012.01.007.

12 Monazam, E.R., Breault, R.W. and Shadle, L.J. (2019) Operational Performance Characteristics of an Aerated L-Valve Operated within a Closed Circulating Fluidized Bed Loop. *Powder Technology*, **354**. https://doi.org/10.1016/j.powtec.2019.05.044.

13 Chovichien, N., Pipatmanomai, S. and Chungpaibulpatana, S. (2013) Estimate of Solids Circulation Rate through an L-Valve in a CFB Operating at Elevated Temperature. *Powder Technology*, **235**. https://doi.org/10.1016/j.powtec.2012.11.016.

14 Wang, D. and Fan, L.S. (2015) L-Valve Behavior in Circulating Fluidized Beds at High Temperatures for Group D Particles. *Industrial and Engineering Chemistry Research*, **54**. https://doi.org/10.1021/ie503951r.

15 Massaro Sousa, L. and Ferreira, M.C. (2020) Analysis of the Performance of an L-Valve Feeding Spent Coffee Ground Powders into a Circulating Fluidized Bed. *Powder Technology*, **362**. https://doi.org/10.1016/j.powtec.2019.12.010.

16 Yazdanpanah, M.M., Hoteit, A., Forret, A., Delebarre, A. and Gauthier, T. (2011) Experimental Investigations on a Novel Chemical Looping Combustion Configuration. *Oil & Gas Science and Technology – Revue d'IFP Energies nouvelles*, **66**. https://doi.org/10.2516/ogst/2010025.

17 Zsembinszki, G., Sole, A., Barreneche, C., Prieto, C., Fernández, A.I. and Cabeza, L.F. (2018) Review of Reactors with Potential Use in Thermochemical Energy Storage in Concentrated Solar Power Plants. *Energies*, **11**. https://doi.org/10.3390/en11092358.

18 Preisner, N.C. and Linder, M. (2020) A Moving Bed Reactor for Thermochemical Energy Storage Based on Metal Oxides. *Energies*, **13**. https://doi.org/10.3390/en13051232.

19 Farcot, L., le Pierrès, N. and Fourmigué, J.F. (2019) Experimental Investigation of a Moving-Bed Heat Storage Thermochemical Reactor with SrBr2/H2O Couple. *Journal of Energy Storage*, **26**. https://doi.org/10.1016/j.est.2019.101009.

20 Cosquillo Mejia, A., Afflerbach, S., Linder, M. and Schmidt, M. (2020) Experimental Analysis of Encapsulated CaO/Ca(OH)2 Granules as Thermochemical Storage in a Novel Moving Bed Reactor. *Applied Thermal Engineering*, **169**. https://doi.org/10.1016/j.applthermaleng.2020.114961.

21 Schrader, A.J., Schieber, G.L., Ambrosini, A. and Loutzenhiser, P.G. (2020) Experimental Demonstration of a 5 KWth Granular-Flow Reactor for Solar Thermochemical Energy Storage with Aluminum-Doped Calcium Manganite Particles. *Applied Thermal Engineering*, **173**. https://doi.org/10.1016/j.applthermaleng.2020.115257.

22 Randhir, K., Hayes, M., Schimmels, P., Petrasch, J. and Klausner, J. (2022) Zero Carbon Solid-State Rechargeable Redox Fuel for Long Duration and Seasonal Storage. *Joule*, Cell Press, **6**, 2513–2534. https://doi.org/10.1016/j.joule.2022.10.003.

23 Vogel, J., Swift, W.M., McDaniel, M.J., Smith, G.W. and Ziegler, A.A. (1978) TESTING OF A PULSED L-VALVE FEEDER FOR USE IN FLUIDIZED-BED COMBUSTION AND SORBENT REGENERATION UNITS.

24 Sinclair, P.G., Katterson, A.M., Klinzing, G.E. and Bachovchin, D. (1983) An Analytical Study of Dipleg Flow Stability Using a Controlled Non-Mechanical Valve. *Powder Technology*, **36**. https://doi.org/10.1016/0032-5910(83)85010-4.

404e Thermo-Photo-Phoretic Phenomena Under High Solar Irradiance Levels for Chemical Processing of Aerosol Flows

Insha Shaikh, Aysha Melhim, Konstantinos Kakosimos

Powder and particles serve as the key raw form for numerous industrial applications, such as the flame and spray pyrolysis, aerosol synthesis, latent and chemical energy storage. In a few of these established and modern applications, particles are exposed to intense temperature gradients and irradiance levels. On illumination, the particles heat up and move in response to the forces exerted by the temperature and/or photon pressure gradients, referred to as thermophoresis and photophoresis respectively. While the thermophoretic movement of particles were considered when designing heat and mass exchangers, since these can facilitate the formation of undesirable deposits on the reactor walls, the photophoretic were typically ignored as negligible. However, this is not quite true for particles under the irradiance of concentrated solar technologies.

Therefore, this study presents an investigation on the thermo-photo-phoretic phenomena encountered in solar aerosol synthesis and chemical processing of materials. Previous studies support the impact of thermo-photo-phoretic phenomena on the transport and deposition of aerosol particles during aerosol synthesis. It has been shown that thermophoresis can lead to the accumulation of particles near the reactor hot walls, while instead photophoresis could reduce particle deposition on the walls. A mathematical model was developed which accounts for the thermo-photo-phoretic phenomena under high solar irradiance levels, specifically for the synthesis of CaC_2.The model considers spherical particles passing vertically through a transparent tube exposed to concentrated light. The effect of different properties of the particles such as size and composition, sunlight (simulated via Xe-Arc light sources) radiation, temperature gradient, and different flow regimes and fluid's rheological properties were examined. In addition, the model was employed to analyze the reactor design, extract the reaction kinetics and process parameters, and quantify energy consumption. Understanding the effects of thermo-photo-phoretic on aerosol synthesis can lead to the development of an optimized reactor design that can be used to improve the efficiency and sustainability of the process. Finally, in this work the first design of an experimental configuration will be discussed in order to evaluate the mathematical model and investigate further the employed phenomena.

404f A Systematic Study on Improving the Flowability of Poorly Flowing API By Dry Coating

Riya Shinde, James Scicolone, Gerardo Callegari, Fernando Muzzio

Powder feeding being the first unit operation in a continuous manufacturing process in oral solid dose products, controls the content uniformity in the final product. Powders (i.e., Active Pharmaceutical Ingredients, APIs) with particle size below 50 μm exhibit poor flow, cohesive behavior, and adversely affect pharmaceutical operations such as feeding, mixing, and granulation. Silicification has thus been introduced to tackle this problem. In this method, nano silica is dispersed through the API using high shear stresses in a dry blending process. Due to the effect of the high shear stresses. Colloidal silica nanoparticles coat the API larger host particles breaking the large intermolecular cohesive forces that produce the high cohesiveness, low flowability behavior of bulk powders. This, in turn, improves the flowability of the API through the continuous system.

This study specifically concentrates on micronized Acetaminophen (mAPAP) used here as a model for challenging APIs to work with. mAPAP has a particle size of 12 μm (D50), and is known to exhibit high cohesion and very poorly flowability, This study aims to improve the flowability and bulk functions of the API blend, by silicifying it in a continuous process. The continuous operation needs to impart high enough shear strain to produce the de-agglomeration of the silicon dioxide nanoparticles, and for enough time to produce the desired dry coating of the host particles by the coating agent. A Comil was used to produce the high-shear conditions and thus dry coat the API and different strategies were put in place to increase the residence time of the material inside the high shear environment by decreasing the size and even by blocking the mesh in the co-mill. The method implemented in this study has helped improve the flow function coefficient of mAPAP from a very poor flowing to a good flowing material in a process that can be adapted in manufacturing environment and integrated into a semi-continuous process.

404g Application of CFD Simulation for the Development of a High Volume Personal Cyclone Sampler

Dzmitry Misiulia, Göran Lidén, Sergiy Antonyuk

Sampling for the purpose of determining dust exposure in the workplace is always conducted actively by means of samplers. A sampling device is specifically designed to enable the sampling of airborne dust which meets the sampling efficiency curve for one of the health-related dust fractions. For the respirable dust fraction, the most common are colloquially known as miniature cyclone samplers as they operate by employing cyclonic action to separate out particles. They should follow the convention for respirable dust sampling with a 50% cut at 4 microns.

Nowadays there is a variety of personal cyclone samplers for respirable dust fractions which operate at flow rates from 1 LPM up to 10 LPM. A series of limit values are already or will be reduced in the future and therefore there is a need for samplers operating at larger flow rates that can collect the required amount of particles for analysis at lower aerosol concentrations. Therefore the main objective of this work was to develop a cyclone sampler for respirable dust fraction at a flow rate of 20 LPM using CFD simulation.

For aerosol flow modelling the Euler-Lagrange method was applied where the continuous phase is treated in an Eulerian manner, whereas the dispersed phase is treated in a Lagrangian approach. For a proper calculation of the flow field, the Large Eddy Simulation (LES) technique was used. A dynamic Smagorinsky-Lilly model was applied to calculate subgrid-scale turbulence.

The developed model was validated on two cyclone samplers, a 9.5 mm in diameter sampling cyclone of the HD design and a 23 mm in diameter FSP 10 cyclone. For that, the simulated performance characteristics, such as a cut-size, the slope of the penetration curve, and the pressure drop, were compared with the experimental measurements. The simulation results agree well with the experimental data. The deviation in cut-size, the slope of the penetration curve, and the pressure drop between the experimental data and LES simulations of the HD cyclone do not exceed 7%, 18.4%, and 10.3% respectively. Regarding the FSP10 cyclone, the deviations in cyclone performance characteristics are even lower.

Two cyclone designs, the HD design, which was originally developed by Higgins and Dewell, and the GK design, developed by Kenny and Gussmann, were developed for a flow rate of 20 LPM. The velocities in the HD cyclone are higher because of the smaller cyclone diameter. As a result, the mean static pressure drop across the HD cyclone is 25% higher as well. Additionally, not the whole cyclone volume is used for particle separation in the HD cyclone. The vortex is not centralised. It ends on the cone wall and there is a dead zone below where the air velocities are close to zero and particles are separated there only due to gravity. In the GK design, the vortex is centralised, penetrates deep into the grit pot, and ends on its bottom wall. Both cyclones provide a cut-size of about 4 microns. However, the penetration curve of the GK cyclone follows the respirable convention better giving a smaller bias/deviation, especially for particles that are 4.5 microns in size and larger.

404h Relating Bulk Solid Caking Events to Phase Change Events That Happen during Crystallization Processes

Kerry Johanson

Caking causes significant problems in the storage of bulk solids in process vessels. Caking can result in significant blockage in a storage silo requiring millions of dollars to clean out. Caking can result in lump formation where lumps can grow to a size capable to block the outlets of silos and hoppers. Caking can increase the rathole tendency of material in storage containers. Caking can cause entire rail cars to be packed with very hard bulk material, making it impossible to empty them with any reasonable method. Caking can cause consumer problems in bags and packages, creating hard packs and making it impossible to dump a bag or package into a User point-of-use location. It is obvious that caking problems reach all scales of the bulk handling system, including silos, hoppers, super sacks, bags, boxes and even small bottles.

Caking is characterized as a gain of significant bulk strength during storage. It is responsible for many lost production hours and unscheduled downtimes in processes dealing with bulk solids – particularly powders. There are several reasons why caking occurs, and the solution to any caking problem will depend on its root cause. In this paper, we will consider just one cause for caking: the mechanism of a recrystallization method. In this method, moisture zones on the surface of the material move and coalesce toward the contact points between particles, forming a pendule. This moisture dissolves some of the bulk material, thereby creating a saturated solution. Subsequent changes in relative humidity surrounding the particles cause the solution to become supersaturated and crystal formation begins to occur within the pendular liquid volume between particles. Likewise, a change in the temperature of the bulk material can result in a supersaturated solution in the pendular liquid between particles, causing crystal growth.

It is important to not only understand the general cause(s) of caking, but to also understand the dynamics behind the caking process, the magnitude of the caking phenomena, how caking relates to the meso-scale behavior, and the time constants associated with caking. Once these things are understood, only then can practicing engineers and material scientists propose methods to mitigate this problem either by process design or by product design.

For this work, three bulk solid systems were examined to gain data and greater understanding of their caking tendencies. These three bulk solid systems are: iodized table salt, calcium chloride flake, and anhydrous acidic acid. The goal of this work is to find common ground between the caking behaviors of these three bulk systems. The caking tendency was measured using a uniaxial strength tester on materials that were prepared with a known concentration of moisture. Caking is a function of storage time and also a function of consolidation stress, so caking strengths were measured as a function of storage time and consolidation stress. These caking experiments were coupled with observations of crystal growth between adjacent particles to find the link that couples the bulk caking behavior to the meso-scale crystallization behavior local to individual particles.

From the experimental work, it was found that these bulk systems tend to gain strength in accordance with two exponential growth terms. The first strength growth term is proportional to storage time to the 1st power. The second exponential growth term is proportional to storage time to the 3rd power. It was determined that this behavior is consistent with phase growth kinetics in accordance with the Johnson-Mehl-Avrami-Kolmogorov (JMAK) equation. This equation suggests that, within a confined space where crystallization is occurring, the fraction of space that transforms into a crystalline material has kinetics consistent with exponential functions that are proportional to time to the 1st power, 2nd power, and 3rd power – depending on the dimension of the crystals that form. Rod-like crystals correspond to kinetics proportional to time to the 1st power. Plate-like crystals correspond to kinetics proportional to time to the 2nd power. The formation of spherical crystals (or 3D clusters) corresponds to kinetics proportional to time to the 3rd power. The bulk strength tends to have a one-to-one correspondence to exponential crystal growth of time to the 1st and 3rd powers where the 1st power exponential strength growth occurs in a short time frame and the 3rd power exponential growth occurs over a longer time frame. This suggests that crystal rods first form in the pendular volume between particles, followed by 3D structure formation in the pendules during bulk solid caking events. Understanding the magnitude and kinetics of these events can give engineers and designers the information needed to prevent or minimize caking issues.

442a The effect of Magnesium Stearate and Colloidal Silicon Dioxide interaction on binary mixture flowability and picking tendency of market image tablets

Jonathan Cummings, Rohit Kumar, Sam Wagoner, Mark Oliveira, Zelalem Worku, Renato Chiarella

Direct compression of drug product formulations with embossed tooling can result in picking of the resultant tablet face. Magnesium stearate (MgSt) is a widely used lubricant in tablet formulations to reduce particle-surface adhesion during compression. Colloidal silicon dioxide (CSD) is another excipient used in small amounts (< 1.5 wt%) within drug product formulations to improve powder flowability by coating larger particles. Previously it has been identified that the surface coverage provided by CSD can negatively impact the lubrication efficacy of MgSt (*Sabir 2001*), thus increasing the opportunity for picking events to occur. Understanding the MgSt and CSD interaction is critical as tablet picking problems typically occur in later stages of production when a process is scaled up. Therefore, identifying the potential for picking in early development can prevent future yield losses and help quantify minimum and maximum concentrations of MgSt and CSD, respectively. The nature of this interaction and its resultant effect on the mixture blend properties and tablet is investigated as a function of the relative amounts of these two excipients in binary mixtures of lactose monohydrate and microcrystalline cellulose using a single embossed compression tooling. The interaction of CSD and MgSt within the formulation is observed under scanning electron microscopy (SEM) and the surface coverage quantified via SEM-based energy dispersive X-ray spectroscopy (SEM-EDS). The formulation's resistance to flow is measured by ring shear testing (RST.Xs) and dynamic flow testing (FT4). The formulation is then compressed on a compaction simulator and its impact on the take-off force, ejection force, evidence of picking via visual inspection, and tablet tensile strength is analyzed.

Reference: Sabir, A., Evans, B., Jain, S., 2001. Formulation and process optimization to eliminate picking from market image tablets. Int. J. Pharm. 215 (1-2), 123–135.

442b Investigation of Electrostatic Properties of Powders to Improve the Continuous Manufacturing of Pharmaceutical Products

Filip Francqui, Aurelien Neveu, Geoffroy Lumay

Continuous manufacturing is a promising future for the production of pharmaceuticals. The continuous processing of pharmaceutical materials can lead to a drift in the powder properties. Therefore, the powder properties evaluated by the characterization before production can be different from the ones observed at the different stages of the production line. This can lead to a lack of processability, a higher variability and to final products that do not meet the strict regulatory requirements.

One of the main problems encountered in the continuous manufacturing of pharmaceutical products is the electrostatic charge buildup that is inherent to the flow of powders. Indeed, tribocharging mechanisms due to the frictional contact between the particles lead to the continuous buildup of electrostatic charges inside the material. These charges contribute to the global cohesiveness of the material, which is associated with a decrease in its flowability. Furthermore, electrostatically charged powders have a high tendency to stick on the surface, leading to the clogging of pipes and inducing variability in the flow. Finally, electrostatic charges can lead to strong agglomeration drastically changing the powder behavior.

A proper characterization is thus essential to allow efficient optimization of the pharmaceutical formulations to be used in continuous manufacturing processes. However, current standardized powder characterization procedures are usually based on old and simple methods that prevent tackling the complexity arising from continuous manufacturing. Especially, electrostatic properties are not taken into account in the common characterization workflow. In this presentation, a new method allowing to precisely evaluate the tribocharging propensity of powders will be presented (GranuCharge, GranuTools). In addition, the useful insights gathered from electrostatic characterization will be demonstrated through practical case studies.

442c Determining the Dissolution Profile of Tablets from Continuous Pharmaceutical Manufacturing Process By Employing Non-Destructive Analytical Spectroscopic Methods

Divyesh Dobaria, Carlos Ortega-Zuniga, Yleana Lugo, George Oze, Sonia M. Razavi, Fernando Muzzio Sr.

In 2004, the FDA released guidance for the industry introducing process analytical technology (PAT) and the concept of real-time release testing (RTRt). As defined by ICH Q8 [1], real-time release testing (RTRt) is the capability to assess and ensure the quality of a drug product during manufacturing, using process data that usually comprises appropriate combinations of measured material attributes and process controls.

The dissolution profile of tablets is a critical quality attribute (CQA) in pharmaceutical development and manufacturing, which needs to be properly monitored and controlled. To ensure effective RTRt and process control, any oral solid dosage application should tackle this concern, as tablet dissolution measurement through wet chemistry is time-consuming and unsuitable for RTRt.

This study aims to develop a model based on process analytical technology (PAT) measurements, process parameters, and the amount of controlled-release polymer to predict the dissolution profile of individual tablets produced in a direct compaction continuous line. The model uses partial least squares (PLS) regression and spectral preprocessing, and its performance is compared to traditional dissolution testing methods. A design of experiments (DoE) was explored to investigate the effect of blender speed, throughput, compression force, and release-controlling excipient and (active pharmaceutical ingredient) API concentrations on the dissolution profile of tablets.

The results were used to create a dissolution soft-sensor model capable of predicting tablet dissolution as a function of the measurable inputs. This model can subsequently be employed to determine, in real-time, whether tablets meet or fail to meet dissolution specifications and devise a diversion strategy for non-compliant products. This strategy can be used to implement real-time quality control of the final product.

References

[1] ICH, Quality Implementation Working Group on Q8, Q9, and Q10 Questions and Answers (Nov. 11, 2010), www.ich.org.

[2] Pawar, P., Wang, Y., Keyvan, G., Callegari, G., Cuitino, A. and Muzzio, F., 2016. Enabling real-time release testing by NIR prediction of dissolution of tablets made by continuous direct compression (CDC). *International journal of pharmaceutics*, *512*(1), pp.96-107.

[3] Andrés D. Román-Ospino, Ravendra Singh, Marianthi Ierapetritou, Rohit Ramachandran, Rafael Méndez, Carlos Ortega-Zuñiga, Fernando J. Muzzio, Rodolfo J. Romañach, Near-infrared spectroscopic calibration models for real-time monitoring of powder density, International Journal of Pharmaceutics, Volume 512, Issue 1,2016, Pages 61-74,

442d Exploring Nondestructive Analytical Methods for Characterizing the Effect of Polymers in Pharmaceutical Formulations from Continuous Manufacturing Process

Divyesh Dobaria, Carlos Ortega-Zuniga, Yleana Lugo, George Oze, Sonia M. Razavi, Fernando Muzzio Sr.

In the Pharmaceutical industry, Critical Quality Attributes (CQA) play a pivotal role in ensuring the efficacy and safety of the final product. It is imperative to evaluate these parameters to get an overall estimation pertaining to tablet properties. Non-destructive methods have emerged as a valuable tool for evaluating tablet properties without altering or damaging the sample. In this study, we focus on how non-destructive methods provide information about tablet properties such as the amount of polymer, tablet hardness, density, and tensile strength by characterizing the effect of polymers in pharmaceutical formulations produced through continuous manufacturing processes. The objective was to create a model that utilized powder blends containing different concentrations of Active Pharmaceutical Ingredients (API) and Polymer. Powder blends were made in V-Blender with an Intensifier bar by varying concentrations of API and excipients but with an identical operation setup. Parameters like varied Blender RPM and Feed Frame RPM were subjected to change in the continuous line. It involved the exploration of the use of various nondestructive techniques, including near-infrared spectroscopy (NIRS), to characterize the effect of polymers in pharmaceutical formulations manufactured using a continuous manufacturing process. The model employed Principal Component Analysis (PCA), Partial Least Squares (PLS) regression, and spectral preprocessing for analyzing the spectral data obtained. These are useful techniques for reducing dimensionality, modeling relationships between variables, and preparing spectroscopic data for analysis. It provides the quantitative and qualitative estimation of the in-line powder blends being characterized. The results obtained from these techniques were compared to traditional destructive methods and UV spectroscopy. The use of these nondestructive methods not only provides a more efficient means of analysis but also has the potential to enhance process understanding and improve product quality control in continuous manufacturing processes.

References

1. Adriluz Sánchez-Paternina, Pedro Martínez-Cartagena, Jingzhe Li, James Scicolone, Ravendra Singh, Yleana C. Lugo, Rodolfo J Romañach, Fernando J. Muzzio, Andrés D. Román-Ospino, Residence time distribution as a traceability method for lot changes in a pharmaceutical continuous manufacturing system, International Journal of Pharmaceutics, Volume 611,2022,

2. Andrés D. Román-Ospino, Ravendra Singh, Marianthi Ierapetritou, Rohit Ramachandran, Rafael Méndez, Carlos Ortega-Zuñiga, Fernando J. Muzzio, Rodolfo J. Romañach, Near-infrared spectroscopic calibration models for real-time monitoring of powder density, International Journal of Pharmaceutics, Volume 512, Issue 1,2016, Pages 61-74,

442e Variation of Spray-Dried Levodopa-Loaded Microcarrier Properties in Correlation with L-Leucine Concentration

Mahasweta Paul, Raymond Lau

To counter the challenges of oral medication of Levodopa in patients of Parkinson's Disease and utilize the advantages of pulmonary drug delivery, inhalable Levodopa formulations with high aerosolization are the need of the hour. The goal of the current work is to evaluate the effects of altering L-leucine concentrations in inhalable co-spray-dried lacto-polymeric Levodopa-loaded microcarriers on their physiochemical and aerosolization properties. Excipients such as a polymer, PLGA (65:35), a carbohydrate, α-Lactose Monohydrate and an amino acid, L-Leucine were selected. The spray-dried Levodopa-loaded microcarriers (LMCs) were then characterized to determine their particle size distribution, drug loading and entrapment efficiency, powder flowability, solid state nature, aerosolization properties and drug dissolution. Increase in L-leucine concentration from 0 to 1% showed a marked improvement in the drug entrapment efficiency (from 17% to 60%), in powder flowability (decrease in CI from 58% to 19%) while the particle size range was observed to decrease. With increasing L-leucine concentrations, PXRD results displayed an increase in the crystallinity of the Levodopa formulations while DSC data indicated an increase in the peak temperature of sublimation from 300°C to 425°C, thereby enhancing their thermostability. At 1% Leucine, aerosolization data showed the highest mean Emitted Dose and Fine Particle Fraction at 79% and 54% respectively. Sustained but similar drug release profiles were observed in simulated lung fluids with a cumulative Levodopa release of upto 85 to 90% achieved by the Hour 19 to 21 for formulations with 0.6% to 1 % Leucine concentration. Overall, it can be concluded that the concentration of L-leucine impacts the physicochemical characteristics and the aerosolization properties of inhalable Levodopa powders.

442f Tuning of Solid-State and Particle Size of Pharmaceutical Nano-Cocrystals

Aaron O'Sullivan, Kevin M. Ryan, Luis Padrela

Introduction

The largest hurdle in the development of new pharmaceutical entities revolves around the increasing prevalence of poor solubility. This barricade is now affecting approximately 40% of the top 200 commercial pharmaceuticals and approx. 90% of drugs in the pharmaceutical pipeline [1, 2]. Poor solubility leads to a less pronounced concentration gradient, slow passive diffusion and *in vivo* failure. Many strategies have been employed to overcome this hurdle, all of which present their own unique advantages and disadvantages. Salt formation, which accounts for approx. 50% of drugs currently on the market, is the most popular method for solubility enhancement, but is however limited by the presence of ionizable groups [3]. Cocrystals are another multicomponent solid form allowing for improvements in solubility and dissolution rate, but require an API and coformer with adequate molecular complementarity and a high degree of hydrogen bonding propensity. Amorphization is another method for improving saturation solubility, but the trade-off includes often large reductions in stability. Particle size reduction to the nanoscale often leads to enhanced dissolution rate, leading to a more poignant concentration gradient promoting rapid passive diffusion, but these particles can sometimes be unstable in solid form or in suspension, and can often be difficult to isolate during manufacture [4].

A single strategy to enhance the solubility/dissolution rate of a poorly soluble drug may not be feasible, due to the vast diversity of pharmaceutical entities currently available. As such, it is necessary to develop adequate knowledge on the simultaneous solid-state and particle size control of pharmaceutical compounds to provide pharmaceutical chemists a multitude of avenues to improve drug physiochemical properties. This study seeks to further the knowledge on solid-state and particle size control of multicomponent solid pharmaceuticals, using scalable and continuous methodologies, attractive at commercial scales.

Method

The most recently approved pharmaceutical cocrystal, Seglentis® (celecoxib + tramadol hydrochloride), was employed as a model cocrystal to demonstrate the ability of the supercritical CO_2-assisted nano-spray drying (SASD) process, described previously [5], to control both solid-state and particle size. A design of experiments (DoE) approach was employed here, highlighted in figure 1, which varied three critical process parameters: solution flow rate, solution concentration, and pressure of the atomizing gas. Briefly, a heated, high-pressure coaxial nozzle with a small mixing volume is employed to atomize a mixture of the drug solution and the supercritical CO_2, into a heated drying/precipitation chamber. The produced droplets are dried, and the resulting particles are collected in a wire mesh placed at the outlet at the bottom of the drying chamber.

Results

Solid-State Control

Powder x-ray diffraction (PXRD) and differential scanning calorimetry (DSC) analysis determined that the processing conditions at DoE point 5 were the only set of conditions which generated the stable (co)crystalline form. This is likely due to larger droplets produced during the atomization step due to the higher solution flow rate, which in turn led to a longer time for complete solvent removal, coupled with a low solution concentration (i.e. low levels of supersaturation) which favored the formation of the stable cocrystalline form. All other sets of process conditions led to the formation of (co)amorphous material due to rapid solvent separation from the solid phase.

The Gordon-Taylor equation (equation 1) was employed to aid in determining the molecular nature of the (co)amorphous samples. This generates a value, T_{g12}, a theoretical glass transition temperature for a mixture of amorphous celecoxib and amorphous tramadol hydrochloride, assuming weak or no intermolecular interactions. Glass transition temperatures exceeding this theoretical value are considered to display evidence of intermolecular interactions [6]. The theoretical value obtained here is 65.1 °C. (Co)Amorphous samples which displayed increased glass transition values in comparison included samples from DoE point 7 (73.9 °C) and 8 (74.0 °C). Further to this, Fourier transform infrared spectroscopy (FTIR) was employed to further investigate these intermolecular interactions. The FTIR spectra displayed in Figure 2 highlights the peak at 3479.9 cm^{-1} which corresponds to the hydrogen bonding network observed in the cocrystal. Unlike other (co)amorphous samples, DoE 8 samples display a peak at this point, indicating hydrogen bonding in the (co)amorphous sample, comparable with the cocrystal.

Particle Size Control

Figure 3 displays SEM images for samples obtained from DoE points 1, 5, 7 and 8. The cocrystalline sample (DoE point 5) displayed a granular morphology and a mean particle size of 1160 ± 220 nm. All (co)amorphous samples were smaller, with a more uniform and spherical morphology. DoE point 4 displayed the smallest particle size of 120 ± 20 nm, due to the low solution flow rate which produces smaller droplets, and the lower solution concentration leading to a lower solid content in the atomized droplets.

Tabletting

Samples from DoE process conditions 1, 5, 7 and 8 were subjected to tabletting (10% w/w sample) and assessed on the basis on tabletability, compressibility and compactability, with fitted tabletting parameters highlighted in Table 1. Samples from DoE 7 displayed the highest tensile strength, with DoE 5 samples displaying the lowest. This may be attributed to the reduced intermolecular interactions in the (co)amorphous samples leading to more efficient packing, however, all samples displayed a tensile strength > 2MPa, which is required for acceptable manufacturability and performance. Formulations including samples from DoE point 5 displayed reduced compactability in comparison to other formulations, as observed in the σ_{t0} value of 5.053 MPa, which may be attributed to the samples larger, more irregular shape. The b value observed in Table 1 provides insight into bonding capacity. Samples from DoE 8 has the largest b value due to its uniform, spherical shape (unlike DoE 5), and its increased intermolecular interactions in unlike samples from DoE 1 and 7. Regarding compressibility, large values for the Heckle coefficient, *k,* indicate good compressibility, with samples from DoE 5 and 8 displaying the best compressibility.

Polymorphic Cocrystals

Due to the production of (co)amorphous (unstable) samples from many of the above tested SASD process conditions, three polymorphic cocrystals were selected to further delve into the solid-state control aspect of this study. These cocrystals are:

- Ethenzamide-Saccharin (EthSacc) (with two polymorphic forms reported) [7]

- Ethenzamide-Gentisic acid (EthGa) (with three polymorphic forms reported) [8]

- Isonicotinamide-Citric acid (IsoCa) (with three polymorphic cocrystal forms and one salt form reported) [9]

A two factorial DoE was employed for the production of these cocrystal systems using the previously described supercritical method, varying solution flow rate and solution concentration. Experimentation has been carried out using SASD, and is ongoing regarding the use of conventional spray drying, and using compressed nitrogen in the SASD apparatus. Briefly, the SASD method led to the production of the stable α and β polymorphs of the IsoCa cocrystal. The SASD method only produced the metastable form II for the EthGa cocrystal, while both the metastable form II, and mixtures of form I and II were produced for the EthSacc cocrystal. Further characterization is ongoing.

Conclusion

The supercritical CO_2-assisted nano-spray drying method described here has been successful in controlling both particle size and solid-state of the most recently FDA approved cocrystal, comprised of celecoxib and tramadol hydrochloride. Variations within acceptable ranges were observed in tabletting in regard to tabletability, compactability and compressibility, demonstrating the methods versatility in manipulating solid-state and particle size to achieve the desired rheological properties. While experimentation and characterization are still ongoing, the supercritical method has shown promise in isolating metastable and stable polymorphs of polymorphic cocrystals, demonstrating the robustness and transferability of this technique to other multicomponent pharmaceutical systems.

References

1. Sathisaran, I. and S.V. Dalvi, *Engineering Cocrystals of Poorly Water-Soluble Drugs to Enhance Dissolution in Aqueous Medium.* Pharmaceutics, 2018. **10**(3): p. 108.

2. Rodriguez-Aller, M., et al., *Strategies for formulating and delivering poorly water-soluble drugs.* Journal of Drug Delivery Science and Technology, 2015. **30**: p. 342-351.

3. Gupta, D., et al., *Salts of therapeutic agents: chemical, physicochemical, and biological considerations.* Molecules, 2018. **23**(7): p. 1719.

4. O'Sullivan, A., et al., *Solid-State and Particle Size Control of Pharmaceutical Cocrystals using Atomization-Based Techniques.* International Journal of Pharmaceutics, 2022: p. 121798.

5. Long, B., et al., *Controlling Polymorphism of Carbamazepine Nanoparticles in a Continuous Supercritical-CO2-Assisted Spray Drying Process.* Crystal Growth & Design, 2019. **19**(7): p. 3755-3767.

6. An, L., et al., *Effects of molecular weight and interaction parameter on the glass transition temperature of polystyrene mixtures and its blends with polystyrene/poly (2, 6-dimethyl-p-phenylene oxide).* European polymer journal, 1997. **33**(9): p. 1523-1528.

7. Aitipamula, S., P.S. Chow, and R.B. Tan, *Dimorphs of a 1: 1 cocrystal of ethenzamide and saccharin: solid-state grinding methods result in metastable polymorph.* CrystEngComm, 2009. **11**(5): p. 889-895.

8. Aitipamula, S., P.S. Chow, and R.B. Tan, *Trimorphs of a pharmaceutical cocrystal involving two active pharmaceutical ingredients: potential relevance to combination drugs.* CrystEngComm, 2009. **11**(9): p. 1823-1827.

9. Stainton, P., et al., *First comparative study of the three polymorphs of bis (isonicotinamide) citric acid cocrystals and the concomitant salt 4-carbamoylpyridinium citrate isonicotinamide.* Crystal Growth & Design, 2018. **18**(7): p. 4150-4159.

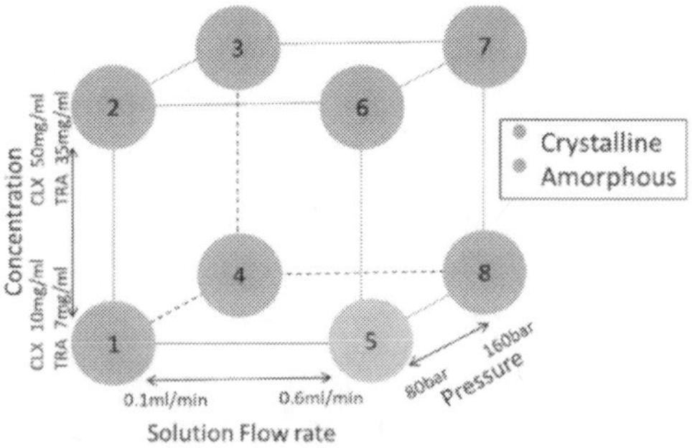

***Fig. 1.** Design of experiments (DoE) cube.*

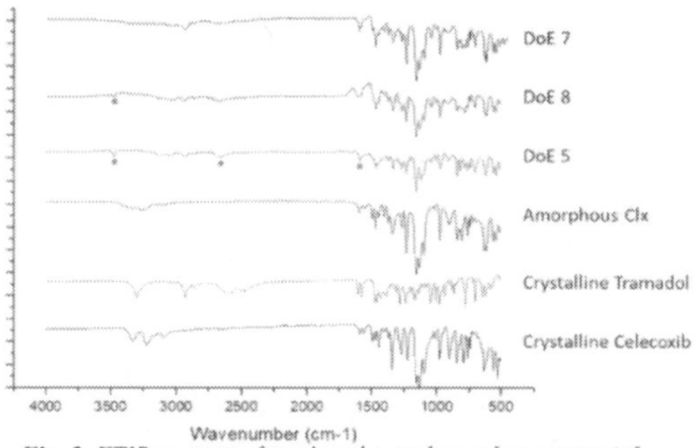

***Fig. 2.** FTIR spectra of produced samples and raw material.*

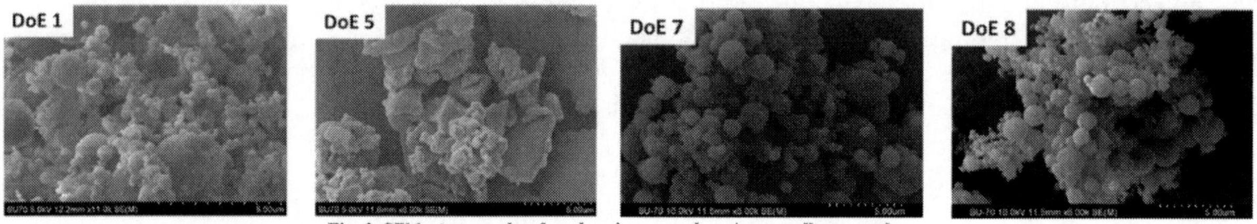

***Fig. 3.** SEM micrographs of produced coamorphous/cocrystalline samples.*

***Table 1.** Fitted tabletting parameters.*

	Tabletability		Compatibility			Compressibility		
Sample	C_p *(MPa)*	R^2	σ_{t0} *(MPa)*	- *b*	R^2	*k (kN)*	P_y *(kN)*	R^2
DoE 1	0.0366	0.9935	17.384	7.96	0.9799	0.33	3.03	0.9388
DoE 5	0.0346	0.9974	5.55	7.91	0.9933	1.11	0.90	0.9071
DoE 7	0.0428	0.9876	17.825	8.64	0.9838	0.33	3.03	0.9472
DoE 8	0.0414	0.9992	14.008	36.7	0.9990	0.60	1.67	0.9926

***Equation 1.** Gordon-Taylor equation.*

$$T_{g12} = \frac{\omega_1 T_{g1} + K.\omega_2 T_{g2}}{\omega_1 + K.\omega_2}$$

$$K = \frac{\Delta C_{p2}}{\Delta C_{p1}}$$

303a Investigating the Applicable Range of the Coarse-Grained Method for Granular Shear Flow

Shota Yokokawa, Hideya Nakamura, Shuji Ohsaki, Satoru Watano

The discrete element method (DEM) has been extensively utilized for modeling and simulating powder handling processes. In a DEM, unsteady motion of each individual solid particle consisting of bulk powder is computed based on Newton's second law. This results in extremely high computational load, limiting the number of particles that can be handled. Number of the particles that can be handled by DEM is approximately from 1 million to 10 million even with the latest workstations. whereas it is necessary to handle more than 1 billion particles for simulation of actual powder handling processes. To address this issue, the coarse-grained (CG) DEM is attracting much attention. The CG-DEM calculates the powder dynamics by replacing the group of original particles with artificially up-scaled single particles (CG particles), thereby reducing the number of particles and computational load. In the CG-DEM, an appropriate scaling law for the CG particles is critical for accurate simulation of the motion of the original particles.

Various scaling laws have been developed for various powder handling processes. However, most of them were based on processes in which the fluid–particle interaction is dominant, and few CG models had been developed for powder mixing processes in which contact and collision between particles are dominant. We have developed a coarse-grained method for granular shear flow (CGSF)[1-3). In the CGSF, original scaling laws that match not only kinetic energy but also friction damping energy, elastic energy, and viscous damping energy between the group of original particles and CG particles were derived. The CGSF also developed a unique method to express the granular shear flow within the group of original particles by scaling the angular velocity of the CG particle. Validity of the CGSF was demonstrated so far[1-3)], however, issues remain concerning the applicable range of the CGSF. Because the CGSF assumed the ideal granular shear flow for the group of original particles when deriving its scaling laws, there is a certain applicable range. However, it is unclear.

In this study, we investigated the applicable range of the CGSF. We employed a drum mixer, which is common powder mixer, as the simulation system. We performed DEM simulations of powder flow in a drum mixer with various particle velocity gradients (i.e., various intensities of the granular shear flow) by changing friction coefficients. We investigated accuracy of the CGSF by comparing the CG-DEM results with the original DEM results, and examined the range of granular shear flow to which the CGSF is applicable.

Reference

1) H. Nakamura, et al., Coarse-grained discrete element method for granular shear flow, *Chem. Eng. J. Adv.* **4** (2020) 100050, https://doi.org/10.1016/j.ceja.2020.100050.

2) N. Kishida, et al., ' Coarse-grained discrete element simulation of particle flow and mixing in a vertical high-shear mixer ', *Powder Technol.*, (2021), **390**, 1-10, https://doi.org/10.1016/j.powtec.2021.05.028.

3) M. Saruwatari, et al., ' Coarse-grained discrete element method of particle behavior and heat transfer in a rotary kiln ', *Chem. Eng. J.*, (2022), **428**, 130969, https://doi.org/10.1016/j.cej.2021.130969.

303b Continuous Synthesis of Drug Particles through Spray-Drying of Nano-Emulsion

Atsushi Oshiro, Akihiro Hatada, Shuji Ohsaki, Hideya Nakamura, Satoru Watano

In the pharmaceutical field, nanoparticle synthesis is an effective means of improving the water solubility of poorly water-soluble drugs. Spray-drying is one of the methods to synthesize fine drug particles. The particle size obtained by the spray-drying method generally depends on the diameter of the sprayed droplets. The synthesis of metallic nanoparticles with single nanometers from micro-droplets generated with electrostatic spraying has been reported. However, the yield of particles obtained by electrostatic spraying is too small, 0.050-15 mL/h. Although spray-drying using a two-fluid nozzle is desirable owing to its high productivity, the diameter of droplets sprayed by two-fluid nozzles is limited to 4μm, leading to a difficulty of synthesis of nanoparticles.

In this study, the "emulsion spray-drying method" was proposed as a novel method for the nanoparticle synthesis. First, nano-sized oil in water (O/W) emulsions with oil-soluble drugs were prepared. Spraying the O/W emulsions generated nano-sized oil droplets within the water droplets. The nano-sized oil droplets dried then to produce fine particles. In this study, nifedipine (NFD) and polyvinylpyrrolidone were used as model powders. Chloroform and Tween 80 were used as the oil component and the surfactant, respectively. These were mixed and stirred with mechanical external force to generate O/W emulsions. The obtained O/W emulsions were confirmed to have monodisperse fine oil droplets with the size of 165 nm. The O/W emulsions were spray-dried, obtaining the NFD particles with the size of ca. 122 nm. To verify the micritization by the emulsion spray-drying method, chloroform solutions of NFD without emulsion were spray-dried. The NFD particles obtained by spray-drying chloroform solutions were 2.5 μm in diameter, demonstrating that fine particles can be obtained by emulsion spray-drying.

Furthermore, ouzo emulsion, which is the self-emulsification method, was investigated to improve the particle yield. Ouzo emulsions are generated through the mixing of oil, water, and co-soluble surfactants. Co-soluble surfactants are well soluble in both oil and water. Phenytoin (PHT), 1-hexanol, and ethanol were used as model powders, the oil component, and the co-soluble surfactant, respectively. These were just mixed to generate ouzo emulsions. The obtained ouzo emulsions had monodispersed fine oil droplets with the size of 7.06 nm. The ouzo emulsion was spray-dried, obtaining the PHT particles with the size of ca. 300 nm. Here, high yields of 2.2 g/min were achieved compared with spray-drying of O/W emulsions. 1-hexanol solutions of PHT without emulsion were also spray-dried. The PHT particles obtained by spray drying a 1-hexanol solution were 1 μm in diameter, demonstrating that fine particles can be obtained by spray-drying of ouzo emulsions. Moreover, dissolution tests were performed to evaluate the solubility of the three PHT particles, including the raw powder. The PHT particles obtained via the spray-drying of ouzo emulsions were confirmed to indicate the highest dissolution rate.

In summary, the spray-drying of O/W emulsions with chloroform and ouzo emulsions with 1-hexanol demonstrated the production of fine drug particles.

303c Dry Granulation of Cathode Active Material and Solid Electrolyte Powders for All-Solid-State Batteries

Shota Miyake, Hideya Nakamura, Shuji Ohsaki, Satoru Watano

Transition from gasoline-powered automobiles to electric vehicles (EVs) has progressed to realize a carbon neutralization. The prerequisite for EVs is the advanced secondary batteries which are safe, high-capacity, and capable of faster charging/discharging. All-solid-state batteries are attracting attention as the next-generation batteries that can fulfill these requirements. As a manufacturing process for all-solid-state batteries, a low-cost and environmentally-friendly dry process, which is solvent-free and does not require drying, is desirable. However, as original materials for all-solid-state batteries are fine cohesive powders, it is necessary to improve their flowability by means of a dry granulation process. For dry granulation, selection and handling of binder particles is of utmost importance. Poly(1,1,2,2-tetrafluoroethylene) (PTFE) is a promising candidate of solid binder particles, because it has been utilized for dry sheeting of fine particles for battery electrodes [1]. However, the fundamental phenomenon of particle agglomeration by PTFE remains unclear.

In this study, we investigated the dry granulation of battery particles using PTFE solid binder in three types of dry granulation processes, which were horizontal high-shear mixer (HSM), planetary ball mill (PBM), and twin-screw hot kneader (TSHK). Generally, in these processes, the dominant external forces exerting the powder are as follows: shear force in the HSM; shear and compressive forces in the PBM; shear and compressive forces with heat in the TSHK. By comparing these different granulation processes, we clarified the requirements for dry granulation with PTFE.

We measured mass fraction of ungranulated fine particles within the granules prepared from three types of granulation processes. It was found that in the case of the TSHK mass fraction of ungranulated fine particles remarkably decreased and the PTFE solid particles were more likely to disperse by transforming to fibrous morphology, leading to agglomerating the fine original particles. This result revealed that by exerting the compressive and shear forces under high temperatures, the PTFE solid binder promotes granulation of the original fine battery particles. Moreover, with an increase in the granulation temperature in the TSHK, the size of granules significantly increased. This was caused by an increasing the number of PTFE fibers and decreasing the diameter of the PTFE fibers under higher temperatures. We then fabricated the electrode sheets by a roller compactor from granules prepared using the TSHK. The tensile strength of the sheets exhibited sufficient strength as compared to that in reference [1]. Furthermore, the sheets fabricated from granules prepared at higher temperatures showed higher tensile strength. We demonstrate that the TSHK process can be employed for the continuous production of electrode sheets for all-solid-state batteries.

[1] F. Hippauf et al, Energy Storage Materials, 21, 390–398 (2019)

303d Generation of Amorphous Dispersions through a Continuous Manufacturing Train to Enable Direct Compression of Reduced Tablet Size Amorphous Dosage Units

Anagha Chandra, Derek Frank, Haichen Nie, Alexander Coelho, Chad Dalton, Hannah Dvorak, Ahmed Elkhabaz, Mairead Fahy, James D. Ormes, Andrew S. Parker, Ashish Punia, Jasmine Rowe, Luke Schenck, Daniel Smith, Neil A. Stroman, Michael Wang, Laura Wareham

Amorphous solid dispersions (ASDs) are an attractive option to improve the bioavailability of poorly water-soluble compounds. However, the material attributes of ASDs can present formulation and processability challenges, which are often mitigated by the addition of excipients albeit at the expense of tablet size. In this work, a continuous ASD manufacturing train combining co-precipitation and thin film evaporation (TFE) was used to generate high bulk-density co-precipitated amorphous dispersion (cPAD). The cPAD/TFE material was directly compressed into tablets at amorphous solid dispersion loadings up to 89 wt%, representing a greater than 60% reduction in tablet size relative to formulated tablets containing spray dried intermediate (SDI). This high ASD loading was possible due to densification of the amorphous dispersion during drying by TFE. Pharmacokinetic performance of the TFE-isolated, co-precipitated dispersion was shown to be equivalent to an SDI formulation. These data highlight the downstream advantages of this novel ASD manufacturing pathway to facilitate reduced tablet size via high ASD loading in directly compressed tablets.

303e Numerical Analysis on Interparticle Adhesion Reduction Mechanism By Surface Modification

Yo Kajiwara, Shuji Ohsaki, Hideya Nakamura, Satoru Watano

Fine particles are used in a wide range of industrial fields, however, handling and processing of fine particles (smaller than ~30 μm) is a generic industrial problem because of their poor flowability. To improve the flowability of fine particles, addition of a flow agent is one of the effective countermeasures. For example, the surface modified particle in which cohesive "host" particle is coated with "guests" flow agents can improve the flowability of host particle. However, many factors such as Van der Waals attraction, Coulomb interaction, etc. affect the flowability, and the detailed mechanism of the flowability improvement by particle surface coating has not been sufficiently clarified. In addition, most studies on the analysis of cohesive force between coated particles focused on the particle static state, not on the dynamic state.

In this study, DEM simulations of coated particles with different coating conditions (number of guest particles on a single host particle, guest and host particle size ratio, different stiffness) were conducted with considering Van der Waals attraction.

Secondly, numerical simulations in which two coated particles collided at constant initial velocity were performed under various coating conditions. In these simulations, cohesive force from JKR theory and Hertzian contact force generated on the contact were calculated. It was found that the cohesive force between the coated particles strongly depended on the contact probability between guest and guest.

Finally, numerical simulation of bulk particles in a rotating drum granulator were conducted and cohesive force and Hertzian contact force between the coated particles were calculated.

It was found that an increase of the contact probability between guest and guest contributed to a decrease in particle adhesion and the improvement of the flowability of the bulk powder.

From these findings, a novel method to estimate the cohesive force between coated particles from the contact probability between guest and guest was proposed with considering the coating conditions such as number of guests, guest and host size ratio and stiffness.

303f Flowability Improvement of Metal-Organic Framework (MOF) Particles Via Pressure Swing Granulation with Small Amount of Binder

Shintaro Yamada, Shuji Ohsaki, Hideya Nakamura, Satoru Watano

Soft metal-organic frameworks (soft-MOFs) are new class of porous crystals formed through coordination bonding of metal ions and organic ligands. Compared to conventional porous materials, soft-MOFs have regularly arranged micropores, large specific surface area, and structural flexibility. The most important feature of soft-MOFs is their adsorption behavior induced by the structural transition. The soft-MOFs show almost no adsorption at low pressure, while the amount of adsorption increases rapidly at a certain pressure. Taking advantage of this property, soft-MOFs are expected to have a wide range of applications, such as adsorbents, separators, catalysts, and sensors. However, the high adhesiveness and poor flowability of synthesized soft-MOFs hinder the industrial application.

One of approaches to overcome this problem is to establish a granulation process for soft-MOF particles. Granulation process is expected to improve flowability and reduce pressure loss in the particle packing layer. Our group has investigated the wet granulation with binder using agitator granulators and fluidized bed granulators. However, previous studies have suggested that the presence of binders in the granules reduced adsorbed amounts in soft-MOFs due to weight of binders. Also, the external mechanical force generated during the granulation process led to a significant decrease in adsorbed amounts. The formation of agglomerate structures with low bulk density, i.e., porous structures, through the binder-less granulation is required to address the above issues.

This study focused on the pressure swing granulation (PSG) method. In the PSG method, compressed gas is injected into the powder layer in the fluidized bed from above and below. The agglomerates then are formed by repeated crushing, fluidization, and consolidation. The van der Waals forces, which are universally possessed by fine powders, and electrostatic forces associated with particle charging should contribute to PSG. The PSG method utilizes the adhesion properties of fine powders, which allows the granulation of fine particles with a small amount of binder and few external mechanical forces. It is also expected that agglomerates can be formed while maintaining the adsorption performance of the original powder. The PSG method has a lot of operating parameters, such as cross-sectional flow velocity, consolidation pressure, crushing pressure, flow time, consolidation time, crushing time, and oscillating pressure. However, it is still insufficient to investigate which parameters affect granule properties.

In this study, therefore, as a preliminary study for soft-MOFs granulation experiments, inexpensive calcium carbonate particles was granulated with PSG method to investigate the effects of operating parameters on the PSG process. Then, utilizing the knowledge obtained from calcium carbonate granulations, ZIF-8, a typical soft-MOF, was granulated with PSG method to evaluate adsorption properties and flowability.

This study revealed that the flowability of the granules improved with an increase in consolidation pressure and the binder concentration. It was also suggested that the too high vibration pressure promoted the crushing of agglomerates, decreasing the flowability of the granules. The number of consolidation-crushing-flow cycles, was also a major factor to improve the flowability. Moreover, based on the operating conditions in calcium carbonate granulation, ZIF-8 particles were successfully granulated with PSG method while maintaining the adsorption amount.

303g Nanoparticles in Motion: The Effect of Gastrointestinal Fluid Chemical Composition on Nanoparticle Stability and Aggregation

Yael del Carmen Suárez López, Alexandra Teleki

Introduction

The gastrointestinal tract (GIT) is a complex system that poses a challenge for the oral delivery of nanoparticles. The different compositions, pH values, and characteristics of each stage of the GIT (e.g., stomach, small intestine, and colon) can vary extensively. Commonly used bulk techniques to assess the fate of nanoparticles through the GIT have been mostly optimized for testing drug compounds' solubility or stability, rather than the dynamic behavior of nanoparticles in the GIT, such as aggregation or enzyme corona formation [1]. Furthermore, little is known about how GIT pathologies may affect nanoparticle stability and physicochemical properties after oral administration. For example, it has been reported that normal pH values, osmolality, and lecithin concentration in the GIT can dramatically change when patients suffer from inflammatory bowel disease (IBD) [2]. These changes should be considered when developing nanoparticles for the oral route. Miniaturization of a system where the GIT is mimicked would offer a high surface-to-volume ratio, the possibility of automatization, and a reduced need for valuable chemicals (like enzymes). Cell-free microfluidic devices that mimic drug compound digestion through the GIT have been previously reported [3], however, this process has not been adapted for the study of nanoparticles yet.

The aim of this work is to develop a microfluidic device in which gastrointestinal media (healthy, and IBD simulating media) can be mixed with nanoparticles, to study how the physicochemical properties of the media affect the stability and aggregation of inorganic nanoparticles through each stage of the GIT.

Methods:

Iron oxide nanoparticles were synthesized by flame spray pyrolysis and coated with a silica layer in a single step. They were characterized by X-ray diffraction (XRD), transmission electron microscopy (TEM), Fourier-transform infrared spectroscopy (FTIR), and energy dispersive spectroscopy (EDS). Commercially available micromixers were then tested. The efficiency with which these micromixers combine and distribute two convergent flow streams to create a uniform mixture was studied. For this purpose, fluorescein dextran MW 10kDa (FITC) was supplied in one flow stream inlet, and water was supplied in another inlet at equal flow rate. The fluorophore intensity throughout the chips at different inlet flow rates (10, 40, and 80 μL/min) was analyzed. Inflammatory bowel disease-simulating media was developed based on literature, and the herringbone micromixers were utilized to study how the pH, bile acid concentration, lecithin concentration, and soluble protein content of healthy and IBD-simulating media affect nanoparticle stability and aggregation behavior.

Results and Discussion

This work presents advances in the design of a microfluidic device based on reported digestion-on-a-chip systems [3]. A pearl-chain micromixer, two types of herringbone micromixers, and a serpentine micromixer were tested for their mixing efficiency. Mixing of FITC was quantified based on pixel intensity in a cross-section of the last part of the microfluidic channel. The fluorescent images obtained were analyzed using a python code available at https://github.com/yaelsuarez/mixing_quantification. The long herringbone micromixer was selected for further development since it presented the best mixing efficiency. Iron oxide nanoparticles coated with silica were characterized by XRD and showed a maghemite/magnetite crystalline phase. FTIR, TEM, and EDS also confirmed the presence of the silica layer coating the iron oxide nanoparticles. The effects of the media composition on nanoparticle aggregation behavior were investigated using the herringbone micromixer to simulate each stage (stomach, small intestine, and colon) of the GIT.

Conclusion

The most efficient micromixer was selected to continue the development of a microfluidic device that allows the systematic study of the different parameters affecting the stability of nanoparticles through each stage of the GIT (stomach, small intestine, and colon), in a healthy or IBD condition, in an automated and controlled way.

Acknowledgments

This project has received funding from the European Research Council under the European Union's Horizon 2020 research and innovation program (grant agreement no. 101002582).

References

[1] S. Asad, Jacobsen, A.C., and A. Teleki. Current Opinion in Chemical Engineering 38 (2022)

[2] C.H. Kim, Y.U. Lee, K.H. S. Kang, G.H. Kang, H. Chu, and S. Lee. Diagnostics 12, 1969 (2022)

[3] P. de Haan, M. A. Ianovska, K. Mathwig, G. A. A. van Lieshout, V. Triantis, H. Bouwmeester, and E. Verpoorte. Lab on a Chip 19 9 (2019)

303i Hydrogen Production from Acid Gas: Structural Changes for Enhanced Reactivity in FeS-Based Hydrogen Sulfide Decomposition Systems

Sonu Kumar, Tanay Jawdekar, Zhuo Cheng, Amanda Trout, Jason Hu, Liang-Shih Fan

Hydrogen Sulfide (H_2S) is a highly toxic gas produced during the processing of fossil fuels. Despite the growing use of renewable energy sources, non-renewable hydrocarbon-based fuels are projected to dominate the world's energy production scenario until 2050. Therefore, efficient, and economic treatment of H_2S during fossil use continues to be an industrial concern. Claus process, the current state of the art for H_2S treatment, faces several drawbacks, such as unfavorable reaction thermodynamics, high capital cost, and the requirement of an air separation unit (ASU). Moreover, it converts the 'H' content of H_2S into low-cost H_2O instead of valuable H_2. The valorization of H_2S to produce H_2 can be achieved through a two-step thermochemical process facilitated by a sulfur carrier undergoing 'sulfidation' and 'regeneration' cycles periodically. In the first step– sulfidation, the H_2S in the acid gas stream reacts with a sulfur lean metal sulfide carrier to produce a sulfur-rich carrier and hydrogen. In the second step– regeneration, the sulfur-rich metal sulfide is regenerated at a higher temperature via sulfur uncoupling, generating sulfur which can be recovered later. For the successful realization of this process, developing a high-performance sulfur carrier that exhibits fast reaction kinetics and stable long-term recyclability is vital. Among various metal candidates, iron sulfide (FeS), a low-cost, environmentally benign, and earth-abundant mineral, can be a suitable candidate for the sulfur carrier material. However, the reactivity of FeS towards H_2S is poor, which hinders its potential for effective utilization. To bridge this gap, we structurally modify FeS to improve its reactivity towards H_2S decomposition.

In this work, we report the design of engineered FeS nanoscale sulfur carrier particles embedded in mesoporous silica SBA-15 (FeS@SBA-15). FeS nanoparticles of size ~3 nm were prepared by the wet impregnation technique. For comparison, micro-sized FeS particles impregnated on SiO_2 support (FeS@SiO_2) were also synthesized. Thermogravimetric analyzer (TGA) experiments showed that FeS@SBA-15 maintained a stable performance over 10 sulfidation-regeneration cycles with nearly ~70% improvement in reaction rate compared to FeS@SiO_2. Several solid characterization techniques like Temperature-programmed sulfidation (TPS), X-ray diffraction (XRD), X-ray photoelectron spectroscopy (XPS), nitrogen physisorption, and transmission electron microscopy (TEM) were used to investigate this improvement in the reaction rate. The analysis revealed that the mesoporous, tubular, and ordered structure of SBA-15 provides the FeS nanoparticles with a large surface area and prevents their agglomeration over extended cycles. Further, fixed-bed studies were carried out to evaluate H_2 yield at a larger scale of operation. To corroborate the experimental results and elucidate the effect of particle size on carrier reactivity, atomistic scale density functional theory calculations were performed. The experimental studies in combination with characterization and atomistic calculations confirm the improved reactivity of the engineered particles. The findings from this study provide vital insights into the design of sulfur carriers for two-step H_2S decomposition.

303h Mitigation of Tribocharging in Pharmaceutical Powders Using Surface Modified V-Blenders

Tanu Mehta, Raj Mukherjee, Ami Shah, Tibo Duran, Bodhisattwa Chaudhuri

Purpose:

The pharmaceutical industry involves handling of powders on a large scale for manufacturing of solid dosage forms such as tablets and capsules constituting about 85% of the total dosage forms. During this manufacturing process, powders get electrostatically charged due to numerous particle-particle and particle-equipment wall collisions and that could lead to compromised powder flow and powder uniformity. In this study, a surface modified blender has been used to analyze tribocharging in pharmaceutical powders. The blender has been fabricated using two types of materials, an insulator, and a conductor. The conductor or the metal arm induces charge of opposite polarity to that of the charge induced by the insulator arm and the overall charge on the powder decreases during the tumbling motion of the blender. In addition to that, a DEM numerical model has been developed using Altair EDEM to provide a mechanistic understanding of tribocharging in pharmaceutical powders using different types of V blenders including surface modified V blender.

Methods:

In this research, Ibuprofen was used as the model drug and processed in aluminum, polyvinyl chloride (PVC), stainless steel, surface modified aluminum-PVC (Al-PVC) and surface modified stainless steel-PVC (SS-PVC) blender at 20% RH for different time intervals such as 2, 10, 20, 30 and 40 mins till charge saturation is achieved. After completion of the blending process, the powder is collected directly in a faraday cup connected to a NanoCoulomb Meter to measure the charge (nC). Each experiment was repeated thrice. For the simulations, the V blender geometry was designed using SolidWorks and DEM model was developed using commercial software Altair EDEM.

Results:

50g of ibuprofen was loaded into the V blender which was then rotated at 13 rpm for different time intervals such as 2, 10, 20, 30 and 40 mins at 13 rpm. The charge saturation was observed at 30 minutes and no significant variation in charge was noted after that and therefore considered as the charge saturation point. In addition to that, it was observed that using blender made of metal, negative polarity charge (Fig 1&4) was observed whereas positive polarity charge was observed using insulator blender (Fig 2). The lowest charge magnitude was observed using surface modified blenders and the polarity of charge was negative (Fig 3&5). Similar observations were made from the DEM simulations.

Conclusion:

We have demonstrated that tribocharging could be significantly decreased by using surface modified blenders in comparison to insulator or metal only blenders. Moreover, it was established that careful selection of equipment materials could significantly reduce the electrostatic charging of powders.

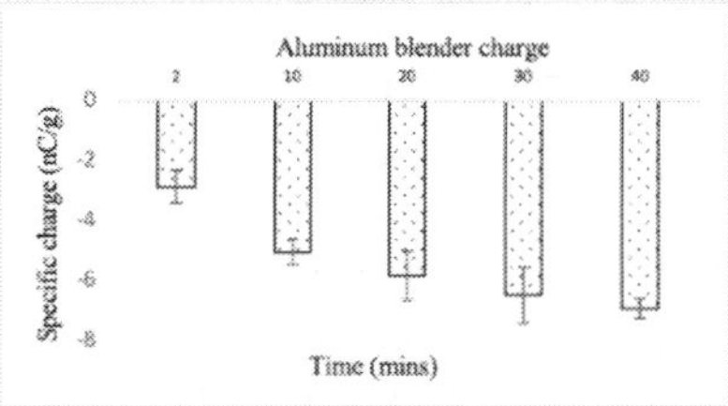

Fig. 1: Shows the specific charge (nC/g) acquired by ibuprofen using Aluminum blender at different time intervals

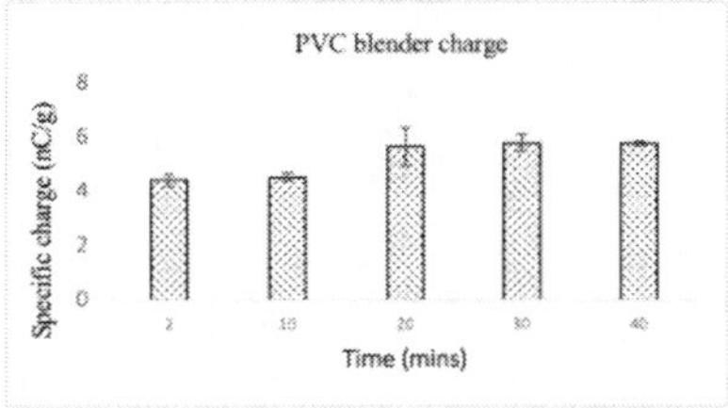

Fig. 2: Shows the specific charge (nC/g) acquired by ibuprofen using PVC blender at different time intervals

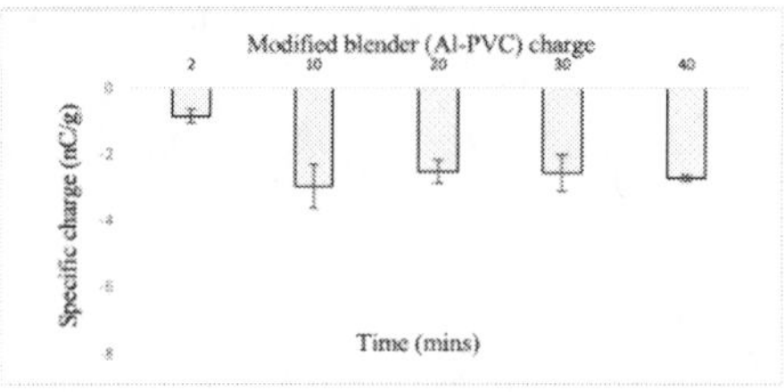

Fig. 3: Shows the specific charge (nC/g) acquired by ibuprofen using modified blender Aluminum – PVC (Al-PVC) blender at different time intervals

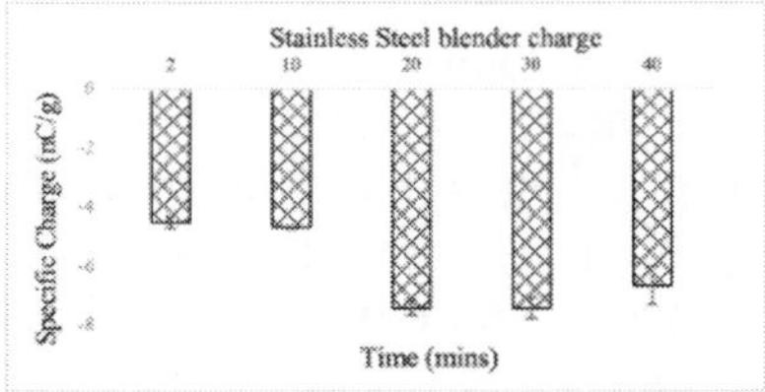

Fig. 4: Shows the specific charge (nC/g) acquired by ibuprofen using stainless steel blender at different time intervals

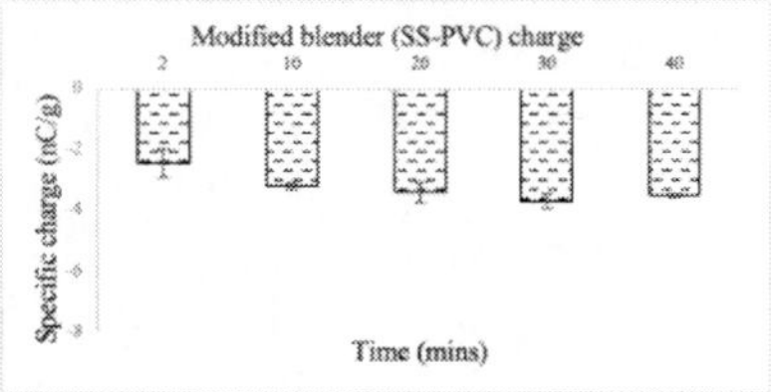

Fig. 5: Shows the specific charge (nC/g) acquired by ibuprofen using modified stainless steel-PVC (SS-PVC) blender at different time intervals

303j Double-Inverse-Opal Structured Assemblies of Titania-Based Photocatalytic Particles Towards Continuous Water Purification Process

Hikaru Namigata, Kanako Watanabe, Keishi Suga, Daisuke Nagao

Photocatalytic decomposition of organic pollutants without using chemical agents is a sustainable way for water purification. Nano–micron sized particulate photocatalysts are commonly employed for an efficient photocatalytic reaction. However, such photocatalytic particles need to be separated from products after the reactions completed, suggesting its limited application to batch reactions. Photocatalytic particles immobilized on supports (e.g., inert inorganic particles or aerogels) or substrates can be easily separated and collected after the reaction, but it decreases the effective surface area of photocatalytic particles. To immobilize or confine the photocatalytic particles without losing their effective surface area, we previously reported the double-inverse-opal (DIO) structured photocatalysts that consist of an inverse-opal silica frame enclosing photocatalytic titania particles in its voids [1]. The voids in the silica frame connected with each other through the interconnecting pores. DIOs exhibited higher photocatalytic activity than the assemblies of titania particles immobilized by a silica frame without voids. This indicated that interconnecting pores have an important role for the reaction.

In this study, photocatalytic activities of DIOs which have ordered or disordered structures were examined. Fabrication of DIOs proceeds by following steps: (i) polymer coating on photocatalytic particles, (ii) self-assembly of polymer-coated particles and infiltration of silica skeleton, and (iii) heat treatment to create voids by removing polymer layers. Monodisperse or polydisperse particles were prepared for the templates for DIOs with different regularity. The ordered DIOs exhibited higher photocatalytic activity, suggesting the importance of regularity of voids to ensure connecting them to each other.

Reference:

[1] H. Namigata *et al.*, *Materials*, **14**, 28 (2021).

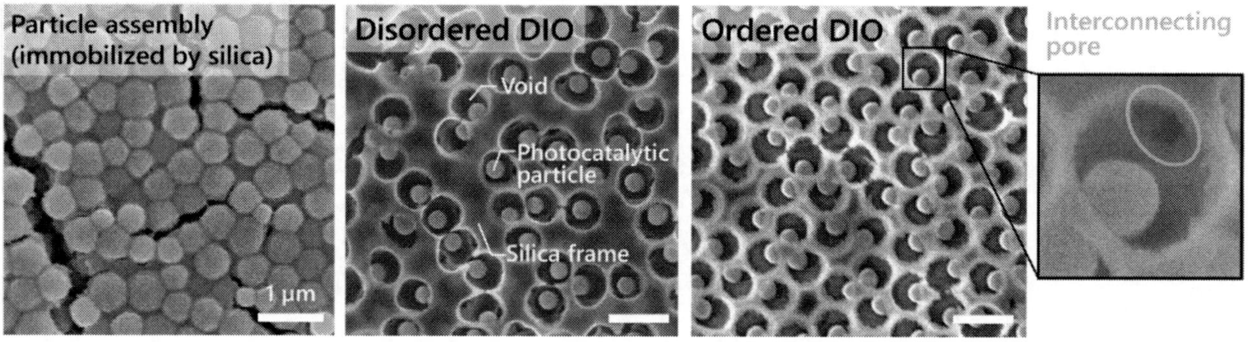

303k Periodontal Biofilm Eradication By Doxycycline Loaded Magnetic Microfibers.

Shaquib Rahman Ansari, Dominique Grimm, Kai Bao, Yael del Carmen Suárez López, Peter Svedlindh, Georgios Sotiriou, Georgios Belibasakis, Nagihan Bostanci, Alexandra Teleki

Periodontitis is a chronic inflammation of the soft tissue surrounding the teeth that leads to the formation of multi-species bacterial biofilms. The disease affects a large portion of the global population and is associated with an increased risk of stroke, hypertension and inflammatory bowel disease. Current treatment of periodontitis is challenging due to the high antibiotic resistance of biofilms, and the difficulty to sustain a sufficient antimicrobial concentration in the oral cavity. Therefore, there is a need to identify novel treatment strategies capable of eradicating established biofilms. Magnetic hyperthermia induced by superparamagnetic iron oxide nanoparticles (SPIONs) has been demonstrated to be effective against bacterial infection.[1] The thermogenic bactericidal approach can facilitate an increased uptake of antibiotics to the bacterial cells and also overcome the limitations of chemical biocides. This study presents a potential local treatment of periodontitis using microfiber composite, which induces local magnetic hyperthermia and antibiotic release for a synergistic biofilm treatment.

Manganese ferrite nanoparticles ($Mn_{0.5}Fe_{2.5}O_4$) of size 18 nm were synthesized using flame spray pyrolysis (FSP). These doped SPIONs show greater magnetic hyperthermia compared to undoped SPIONs.[2] Thus, $Mn_{0.5}Fe_{2.5}O_4$ nanoparticles were incorporated into poly(methyl methacrylate) (PMMA) microfibers by electrospinning. Electrospun microfibers were also prepared using poly(lactic-co-glycolic acid) (PLGA) and loaded with doxycycline hyclate (DXY). The PMMA/SPION and PLGA/DXY microfibers were combined into a composite structure and evaluated for their effect on periodontal biofilms.

PMMA/SPION microfibers reached a temperature of 42-43 °C upon exposure to an alternating magnetic field (AMF), at clinically relevant AMF amplitude and frequency. The PLGA/DXY microfiber showed a cumulative release of 35.3% of the loaded drug in the first 5 min of the in vitro drug release study and plateaued at an accumulated dose of 47 μgmL^{-1} after 24 h. The two microfiber materials were successfully joined by compacting them at 65 °C.

Electrospinning was successfully used to produce PMMA/SPION and PLGA/DXY microfibers. The magnetic and drug loaded microfibers show mild hyperthermia and sufficient drug release, respectively. This study demonstrates that DXY loaded magnetic microfibers can be used to eradicate periodontal biofilms.

Acknowledgement:

The authors would like to acknowledge the financial support from the Science for Life Laboratory. This project has received funding from the European Research Council (ERC) under the European Union's Horizon 2020 research and innovation programme (grant agreement No. 101002582).

References:

(1) Alumutairi, L.; Yu, B.; Filka, M.; Nayfach, J.; Kim, M.-H. Mild Magnetic Nanoparticle Hyperthermia Enhances the Susceptibility of Staphylococcus Aureus Biofilm to Antibiotics. *Int. J. Hyperth.*, 2020, *37* (1), 66–75.

(2) Ansari, S. R.; Hempel, N. J.; Asad, S.; Svedlindh, P.; Bergström, C. A. S.; Löbmann, K.; Teleki, A. Hyperthermia-Induced in Situ Drug Amorphization by Superparamagnetic Nanoparticles in Oral Dosage Forms. *ACS Appl. Mater. Interfaces*, 2022, *14* (19), 21978–21988.

3031 On the Relationship of Interparticle Force and Bonding Number: Predicting Breaking Force of Pharmaceutical Powders and Blends

Zhixing Lin, Christopher Kossor, Rajesh Dave

Tablets are a widely used pharmaceutical dosage form, but their formulation development is heavily reliant on empirical design of experiments and previous expert knowledge. This creates a need for predictive and mechanistic computer-based design models that could enable faster formulation development based on material properties and fewer particle-scale experiments. However, existing prediction models have limitations in terms of empirical fitting of models based on powder properties, which may not provide fundamental insights into particle interactions. To address this, we propose a mechanistic model that predicts tablet breaking force/tensile strength for single and binary component formulations based on bonding strength and bonding number, which is the number of particle bonds per area. This model allows for simplified estimation of rearrangement and fragmentation. To estimate the bonding number, we derive an estimation from JKR model and adopt our previously proposed adhesion force model into bonding force estimation. We further use a dimensionless properties averaging approach for predicting the properties of binary component tablets from those of single-component properties. The proposed model shows good correlation with experimental results for different active pharmaceutical ingredient powders, compression pressures, and particle sizes. The experiments used to validate the model include eight single-component and nine binary-component formulations. Our mechanistic model has the potential to simplify the tablet formulation process for binary and multi-component tablets, requiring only properties assessment for individual/single components. This approach could enable faster and more efficient design of tablet formulations in the pharmaceutical industry.

Reference

1. Eriksson M, Alderborn G. The effect of particle fragmentation and deformation on the interparticulate bond formation process during powder compaction. Pharmaceutical research. 1995;12:1031-9.
2. Sun CC. Decoding Powder Tabletability: Roles of Particle Adhesion and Plasticity. Journal of Adhesion Science and Technology. 2011;25(4-5):483-99.
3. Ryshkewitch E. Compression strength of porous sintered alumina and zirconia: 9th communication to ceramography. Journal of the American Ceramic Society. 1953;36(2):65-8.
4. Wu C-Y, Best SM, Bentham AC, Hancock BC, Bonfield W. A simple predictive model for the tensile strength of binary tablets. European Journal of Pharmaceutical Sciences. 2005;25(2-3):331-6.
5. Sebhatu T, Alderborn G. Relationships between the effective interparticulate contact area and the tensile strength of tablets of amorphous and crystalline lactose of varying particle size. European journal of pharmaceutical sciences. 1999;8(4):235-42.
6. Capece M, Huang Z, To D, Aloia M, Muchira C, Davé RN, et al. Prediction of porosity from particle scale interactions: Surface modification of fine cohesive powders. Powder technology. 2014;254:103-13.
7. Johnson KL, Kendall K, Roberts AD. Surface Energy and the Contact of Elastic Solids. Proceedings of the Royal Society A, Mathematical, physical, and engineering sciences. 1971;324(1558):301-13.
8. Capece M, Silva KR, Sunkara D, Strong J, Gao P. On the relationship of inter-particle cohesiveness and bulk powder behavior: Flowability of pharmaceutical powders. International journal of pharmaceutics. 2016;511(1):178-89.

303m Heat Transfer and Flow Characteristics in Internally Circulating Fluidized Bed Lab-Scale and Pilot Scale Facilities

Hideki Takeuchi, Yusumi Nagahashi, Takahiro Oishi, Yutaka Asako, Tetsuo Takeuchi, Shungo Hashioka, Hirokazu Kajiwara, Mikiya Watanabe

Heat transfer on a horizontal heat transfer tube is investigated in internally circulating fluidized beds(ICFB®). Experiments were carried out both in lab-scale and pilot scale facilities. Heat transfer results are compared with results of PIV (Particle Image Velocimetry) analysis of particle movement around heated tube surfaces. In the circulating bed, the heat transfer coefficient was found to be lower than in a bubbling or turbulent fluidized bed. This can be explained from the PIV results which show that the degree of deterioration depends on the movement of particles around the heat transfer tube. Trends and values of heat transfer coefficient were almost the same for the lab-scale and pilot-scale facilities. The results indicate reasonable agreement between the lab-scale and pilot-scale facilities.

303n Advancements in the OSU Moving Bed Chemical Looping Biomass Gasification: >700hrs Sub-Pilot (15kWth) Operation, Modularization, and Cross Current Reactor Design

Rushikesh Joshi, Cody Park, Yaswanth Pottimurthy, Eric Falascino, Dikai Xu, Dawei Wang, Anuj Joshi, Sonu Kumar, Pinak Mohapatra, Qiaochu Zhang, Ashin Sunny, Qichang Meng, Soohwan Hwang, Vedant Shah, Andrew Tong, Liang-Shih Fan

Abatement of the effects of anthropogenic climate change remains one of the most imposing problems confronting the scientific community. Commercializing biomass-based technologies as an alternative to fossil fuels for a sustainable future faces several challenges, including high pretreatment cost, tar formation, high moisture content, process complexity, ash separation, etc.

This current study presents advancements in OSU's biomass-to-syngas (BTS) technology, a robust chemical looping gasification process for producing syngas from biomass. The technology utilizes iron-based oxygen carriers, which donate their lattice oxygen for lignocellulosic biomass gasification. The BTS technology consists of a moving bed reducer, where the biomass is gasified into syngas, and a fluidized bed combustor for regeneration in the air of the oxygen carriers.

Xu et al., in 2018, established the BTS process with a moving bed bench scale reactor which showed high-purity syngas formation with a variable H_2:CO ratio. The current work presents the scale-up results that prove the BTS technology's commercial viability on a sub-pilot scale (~15 kWth capacity) with continuous operation for over 700 hours. The demonstration successfully tested three types of untreated biomass, including pelletized corn cob, woody pellets, and unpelletized loose corn cob, for a prolonged duration with a variable syngas H2/CO ratio quality ~2 and a syngas purity greater than 70%. Process simulations show that integrating the BTS system with the Fischer Tropsch process for liquid fuel production can achieve a ~14% reduction in biomass requirement compared to conventional biomass gasifiers. Heat integration can lead to an overall autothermal process using combustor exhaust heat for biomass drying.

Further bench-scale studies are done to determine the minimum required reactor volume and the minimum enhancer gas flow rate for efficient gasification. A volume reduction of nearly 67% was achieved, with a minimum steam flow rate as low as 5% of the inlet carbon flow rate for woody biomass pellets. An increase in enhancer gas injection leads to higher tar concentrations due to decreased gas residence time in the reactor.

Furthermore, the study introduces a novel cross-current moving bed reducer design, which involves outletting the syngas from the middle of the reactor and inletting the enhancer gas from the bottom. This design significantly improves the char gasification rate and syngas yield, as confirmed by experimental and process simulation results.

The study also proposes a modular system for generating high-purity syngas from biomass and capturing CO_2 by operating two moving bed reducers in parallel with a common fluidized bed combustor. This system eliminates the need for a separate CO_2 capture unit and recovers additional hydrogen/heat, boosting the process's effective thermal efficiency (ETE) by 17%. The modular system can be integrated with other syngas-requiring processes and can recover energy from tail gas streams, producing additional hydrogen or heat output while capturing CO_2.

Finally, the study highlights future research directions, including co-gasification with carbon-rich sources such as post-consumer plastic waste, to enhance syngas quality via experimental results and process simulations.

303o Exploring Magnetically Stabilized Fluidized Bed Reactor for Enhanced Point Source CO_2 Capture Using Alkali Metal Carbonate-Based Solid Sorbents

Ashin Sunny, Ishani Karki Kudva, Pinak Mohapatra, Anuj Joshi, Liang-Shih Fan

Despite the increasing use of renewable resources for power generation, natural gas and other fossil fuels will continue to have a significant role in power generation for the foreseeable future. Unfortunately, fossil fuel combustion results in carbon dioxide (CO_2) emission into the atmosphere, contributing significantly to greenhouse gas emissions. CO_2 levels have increased rapidly since the Industrial Revolution, mainly due to anthropogenic activities, resulting in harmful climate change. This issue is one of the world's most pressing problems today, necessitating the urgent development of reliable, scalable, and cost-effective CO_2 capture technologies for Natural Gas Combined Cycle (NGCC) power plants.

CO_2 capture can be achieved through solvent-based, sorbent-based, or membrane-based approaches. Although solvent-based technologies are currently state-of-the-art for capturing carbon emissions, sorbent-based methods have potential advantages, such as being less energy-intensive, having a wider temperature range, and being easier to handle and dispose of. Alkali metal carbonate-based sorbents are commonly used due to their low cost and high CO_2 sorption capabilities, but there are issues with pressure drops and inefficient contact. Efficient and stable sorbents with good mass and heat transfer characteristics can make solid-sorbent technology competitive with solvent-based methods.

Our previous study investigated using potassium carbonate (K_2CO_3)-based solid sorbent to capture CO_2 from simulated flue gas containing 4% CO_2 and 10% water. The sorbent adsorbs between 50°C to 90°C and can be regenerated at 100°C to 150°C. Metal oxide-based supports and other support materials are used to disperse K_2CO_3 and increase the surface area. The sorbents are characterized using x-ray diffraction (XRD), scanning electron microscopy (SEM), transmission electron microscopy (TEM), nitrogen physisorption and diffuse reflectance infrared fourier transform spectroscopy (DRIFTS). Thermogravimetric analysis (TGA) and fixed bed experiments are conducted to test the material's recyclability, stability, and performance.

To compete with solvent-based methods, not only do we need more efficient and stable sorbents, but also an innovative technology that facilitates efficient contact between gas and solids for good mass and heat transfer. A potential solution is a magnetically stabilized fluidized bed reactor (MSFB) with a magnetically susceptible bed, which can suppress bubble formation and enhance CO_2 capture. This technology stabilizes the fluidized bed and extends the particulate fluidization regime, preventing sorbent elutriation and eliminating the need for a gas-solid separation device or cyclone. This also increases gas-solid contact, simulates a plug flow reactor-type operation, substantially reduces pressure drop, and reduces particle loss due to less particle/particle-wall collision.

In this study, a uniform magnetic field is generated in the bed using a Helmholtz coil and the effectiveness of a magnetically stabilized fluidized bed is tested. Different magnetically susceptible bed materials like ferrimagnetic Fe_3O_4 particles are used to impart magnetic properties to the bed. The pressure drops across the bed are measured with and without applying a magnetic field. Results indicate that lower pressure drops are observed when magnetic fields are applied. Detailed investigations on fluidization behavior and regimes are conducted with a comparison between magnetized and non-magnetized conditions and supplemented with computational fluid dynamics (CFD) simulations incorporating magnetism. K_2CO_3 sorbent from the previous study is impregnated onto different magnetic materials and tested for CO_2 capture performance in a fixed bed reactor. The optimized magnetic sorbent is tested under reaction conditions in a hot MSFB setup, and the results are compared to the non-magnetic K_2CO_3 sorbent by analyzing CO_2 capture efficiency, CO_2 breakthrough curves, pressure drop across the bed, and sorbent entrainment. To de-risk commercialization efforts, the project's next phase involves designing and fabricating a bench-scale unit to study different modes of operation, including fixed bed, bubbling fluidized bed, and magnetically stabilized bed, to quantify and validate MSFB performance.

This research project can significantly contribute to both the carbon capture and fluidization fields by providing scientific and technical insights into the development of solid sorbents and innovative methods like magnetically stabilized fluidized beds for improved capture efficiency and overall CO_2 uptake. Addressing climate change by developing low-cost carbon capture technologies is the need of the hour, and this work is a promising step toward achieving this goal by capturing CO_2 from natural gas-based power plants through enhanced gas-solid contact.

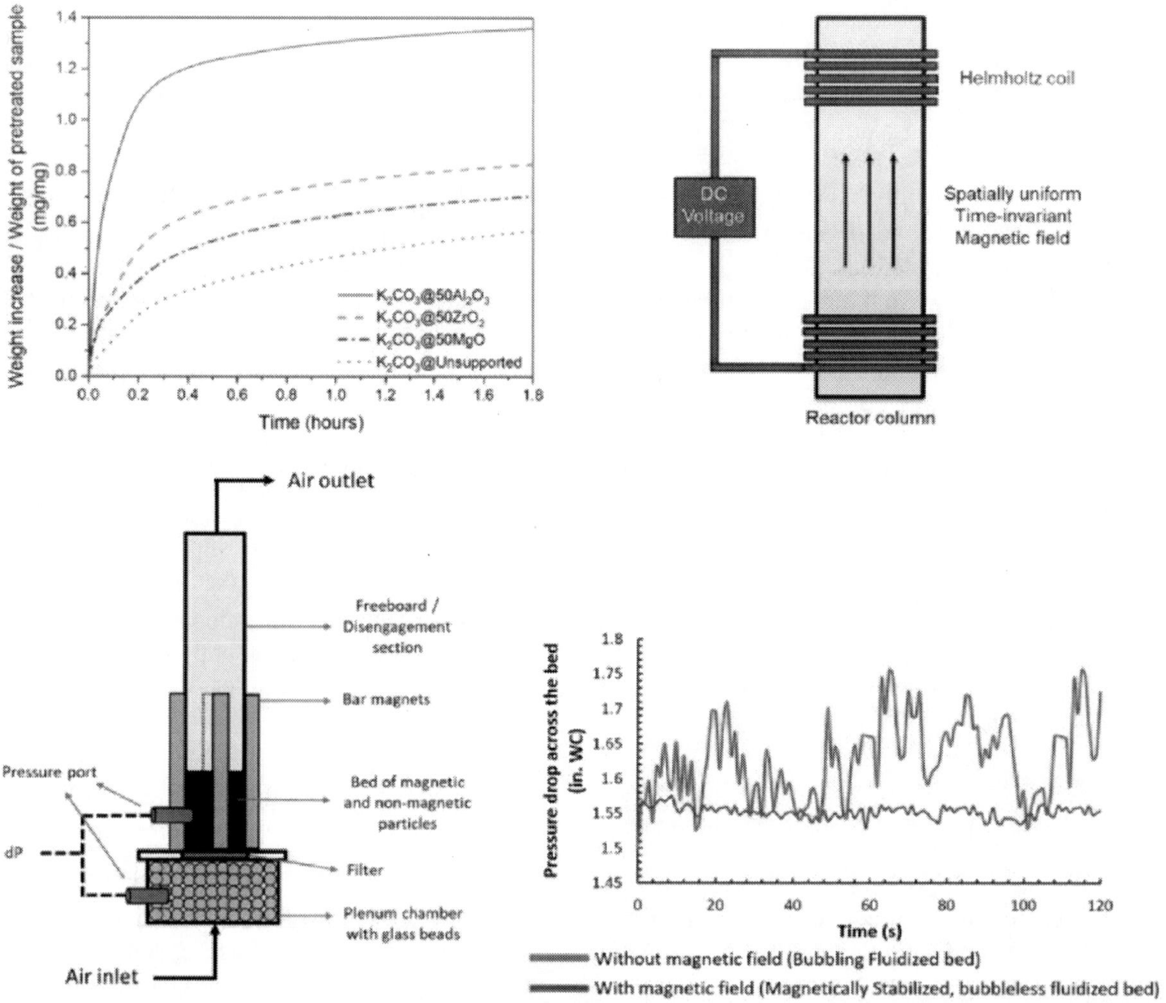

Weight increase in TGA with time for K_2CO_3@Unsupported and K_2CO_3@50(Al_2O_3, ZrO_2, MgO) during CO_2 adsorption (top left), Magnetic bed schematic using Helmholtz coil (top right), Schematic of magnetic bed using bar magnets (bottom left) and pressure drop across the bed with and without magnetization (bottom right)

303p Catalytic Methane Pyrolysis for Hydrogen and Carbon Nanofiber Co-Product

Jessica Hauck, Kent J. Warren, Mija H. Hubler, Boning Wang, Samantha Harshberger, Andrew Broerman, Christopher Gump, Theodore Champ, Alan Weimer

Hydrogen is one of the most essential chemical building blocks to our energy infrastructure, with downstream applications in oil refining as well as production of ammonia, the key component in fertilizer. Today, nearly all hydrogen is produced commercially via steam methane reforming. For every kilogram of hydrogen produced, there are roughly 9 kilograms of CO_2 released to the atmosphere.[1] With an estimated global hydrogen production rate of 75 million metric tons per year[1], over 675 million metric tons of carbon dioxide are emitted via this process. A profitable alternative to produce hydrogen while circumventing greenhouse gas (GHG) is needed. A Catalytic-Chemical Vapor Deposition (CCVD) process to produce hydrogen and carbon nanofibers (CNFs) from methane is a financially promising alternative to steam methane reforming.

Methane, the primary component of natural gas, requires substantial process conditions to overcome the strong C – H bonds. Particle Atomic Layer Deposition (ALD) enables the formation of highly dispersed metal nanocatalysts on a given support which reduce the energy required for methane conversion. Moreover, the highly dispersed catalyst, consisting of abundant metals such as iron or nickel, is essential as this process contains a "sacrificial catalyst"; process contains a *sacrificial catalyst*; the catalyst, support, and CNFs combine as one product to improve the strength and durability of concrete. In partnership with both the Hubler group in the department of Civil and Environmental Engineering and the National Ready Mixed Concrete Association (NRMCA), we are investigating the CNF/catalyst product to reduce cracking and improve the durability of high purity concrete (HPC).

In this research, abundant transition metals are deposited onto a support via particle ALD. CCVD and catalyst treatment parameters are investigated to maximize carbon and hydrogen yield. CNFs and hydrogen are then generated via CCVD with the particle ALD catalyst. A preliminary technoeconomic analysis is performed to assess the large-scale viability of this catalytic methane pyrolysis technology.

[1]Bartlett, J., & Krupnick, A. 2020. "Decarbonized Hydrogen in the US Power and Industrial Sectors: Identifying and Incentivizing Opportunities to Lower Emissions" (Report 20-25). *Resources for the Future*.

303q A Comparative Analysis of the Influence of Macromolecular Templates on the Growth of Calcium Carbonate Crystals Using a Droplet-Based Microfluidic System

Eun-Hye Jang, Bolam Kim, You-Kwan Oh, Il Kim, Sungwook Chung

Biomineralization is the process by which living organisms produce minerals, often in intricate and highly organized structures. Biominerals are complex inorganic-organic composites, and they have various sizes ranging from millimeters to nanometers. One of the most common biominerals is calcium carbonate ($CaCO_3$), which is often found in a wide range of organisms, from marine mollusks to human bones. Living organisms use biomolecular controls to create biominerals exhibiting complex hierarchical structures. The challenge is to link macromolecule-crystal interaction to changes in kinetics, growth mechanism, size, and shape.

The crystal growth of calcium carbonate is a complex process that is influenced by a variety of physical and chemical controls, including temperature, pH, and the presence of other additives in the solution environments. In recent years, there has been increasing interest in studying the crystal growth of calcium carbonate under microfluidic environments, which can provide precise control of the flow and composition of the fluid. The use of microfluidics has the potential to provide new insights into the mechanisms of biomineralization and to facilitate the development of new materials with tailored properties.

Herein, we synthesized a category of polymers, including hyperbranched polymers (HBPs), derivatized with functional groups such as carboxyl acids and amine moieties that have a relatively strong electrostatic interaction with cationic Ca^{2+} and anionic CO_3^{2-} ions. To gain insights into the interactions between crystals and functional macromolecules, we conducted a comparative study on the influence of macromolecular templates on the growth of calcium carbonate crystals using both a bulk and a droplet-based microfluidic system. We obtained information such as nucleation density, crystal number density, and polymorphs of the crystals from growth kinetics via various analytical methods, including *in situ* optical microscopy equipped with a high speed CCD camera. The results of $CaCO_3$ crystal formation mediated by functional macromolecular templates under the bulk and microfluidic environments will be discussed in the presentation.

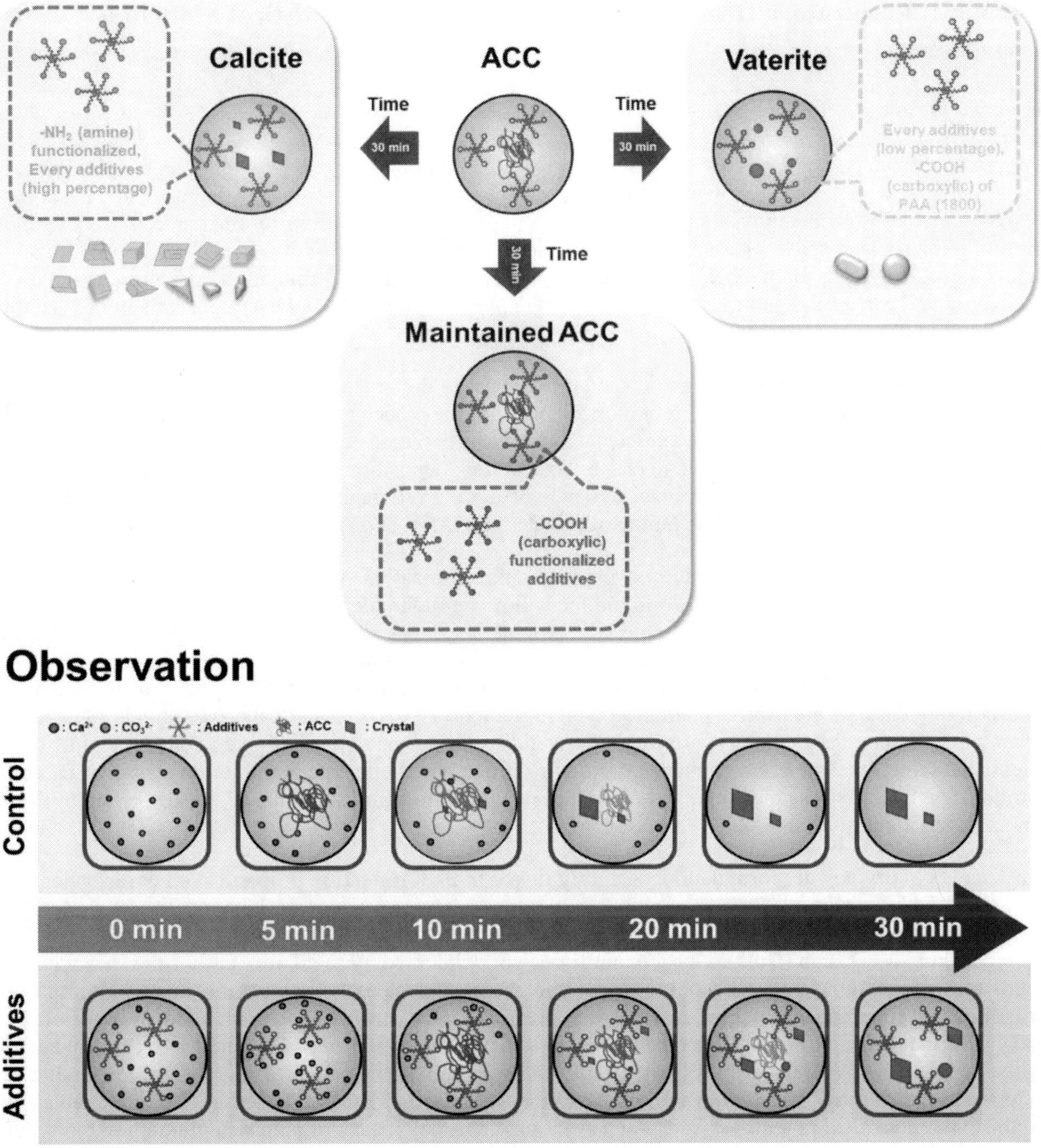

Calcite
ACC
Vaterite
Time
30 min
-NH$_2$ (amine) functionalized, Every additives (high percentage)
Every additives (low percentage), -COOH (carboxylic) of PAA (1800)
Maintained ACC
-COOH (carboxylic) functionalized additives
Observation
: Ca^{2+} : CO$_3^{2-}$: Additives : ACC : Crystal
Control
0 min
5 min
10 min
20 min
30 min
Additives

303r Integrating Three-Fluid Nozzle Technology as a QbD Tool for Enhancing Advanced Respiratory Drug Delivery Formulations: A Statistical Modeling Approach

Mykaella Mestre, Rui Churro Sr., Lídia Santos, Ana Aguiar-Ricardo

Dry powder inhalers, DPIs, have been gaining increased attention as an effective treatment for both local and systemic diseases, owing to their distinct advantages in terms of better physical and chemical stability. Conventional DPI formulations are traditionally composed by physical mixtures of coarse carriers with one (or more) micronized active pharmaceutical ingredients (API), typically with aerodynamic particle size below 5 μm. However, despite being a well-developed approach, these formulations still present some challenges and carrier-free composite particles, where one or more APIs and excipients are embedded into a single solid phase arise as alternative.

This approach improves uniformity and dispersibility, having the potential to increase therapeutic activity. Composite particles can present different morphologies, namely encapsulated particles, where the API core is surrounded by an outer layer of an excipient. The preparation of such particles within the inhalable range requires particle engineering methods to ensure correct lung deposition and increased bioavailability. Among those, spray drying (SD) has been shown to be a preferred method to prepare encapsulated composite particles as it enables fine tuning of powder properties and is an easy-scalable process.

A three-fluid external nozzle (3FN) is a novel pneumatic atomization strategy that features a three-layered concentric structure. In this design, two liquid feeds flow through independent channels, thus overcoming solubility and solvent incompatibility issues and eliminating the need for co-solvent systems that solubilize all components in a single solution such as those often found for traditional atomization strategies (traditional approach). The three-fluid nozzle enables swift and scalable manufacturing of formulations incorporating encapsulated active pharmaceutical ingredients (APIs). This advancement holds significant advantages for lung drug delivery, offering enhanced therapeutic outcomes, targeted delivery, controlled release, and improved patient compliance. Using this process tool one aqueous solution containing a shell-forming agent is fed to the outer channel of the nozzle while an organic solution of the API is fed to the innermost channel.

The success of dry powder formulations is tightly linked to its formulation properties, however, when manufactured by means of spray drying the interrelations between product's critical quality attributes (CQAs) and manufacturing potential critical process parameters (pCPPs) pose a challenging task. On that frame, Quality by Design (QbD), a systematic framework endorsed by regulatory authorities, has gained significant momentum for its potential to streamline pharmaceutical development and manufacturing processes. Embracing the principles of QbD, it is pivotal to have a systematic understanding of the pCPPs and CQAs that can significantly enhance the manufacturing robustness and efficiency. To reach that goal a statistical approach is often employed, by means of design of experiments (DoE), reducing the number of required tests that maximize the interactions between pCPPs and CQAs. A DoE as a QbD tool enables to establish mathematical relationships between cPPs and CQAs by means of empirical modelling tools, such as statistical regression models.

The goal of this study was to employ a QbD strategy to develop empirical models for CQAs of DPIs, consisting of encapsulated composite particles prepared using a 3FN. A DoE was performed where the investigated CQAs were particle size, reported as the Dv50, and aerosolization parameters, namely fine particle fraction FPF and emitted dose, ED. Spray drying parameters such as the outlet temperature (T_out), the atomization flow rate (F_atom) and the ratio between the feed flow of the organic and aqueous solution (M_ratio) were considered as pCPPs, while other parameters were kept constant for a model formulation of trehalose:leucine (80:20 % w/w) containing Itraconazole as model drug.

The DoE was based on a three-factor full factorial design and a central point. An exploratory data analysis (EDA) enabled the identification of relationships and patterns in the data. The empirical models were developed by fitting a multivariate linear regression (MLR) by least squares. Stepwise variable selection was applied to ensure only significant variables were included, thus assuring model simplicity. The stopping criterion for the variable selection was minimization of the Akaike information criterion (AIC). Moreover, outlier identification was performed by Hotelling's T^2 distribution analysis.

The developed models were evaluated with a hypothesis test, whereby the null hypothesis must be rejected, for a given significance level α (α=5% in this study). To evaluate the models' significance an ANOVA analysis was performed. Model fitness to the data was evaluated by the coefficient of determination , and the adjusted coefficient of determination, , as the latter prevents the selection of over-fitted models. Multicollinearity was checked by the variation inflation factor (VIF), considering a VIF threshold of <5. Lastly, to validate the models and assess their predictive capacity, model predictions were compared with a new set of independent experimental data.

From the EDA, linear relationships could be identified, it could also be observed that, likely ED was independent on the studied pCPPs because all tests yielded similar results. Therefore, no regression model was developed for this CQA, as it is likely a cause of the selected formulation without any significant statistical impact of the tested cPPs. Regarding particle size (Dv50), variable selection identified T_out and F_atom as the factors that explained most variability of this CQA, with F_atom's coefficient being higher than for T_out, as seen in Table 1. According to the model increased atomization and temperature yield finer particles. The values of and were 0.86 and 0.82, respectively, demonstrating that the model adequately fits the data. The inclusion of T_out as a predictor was not expected as it does not influence the droplet size directly. Furthermore, no second order interactions were considered. Therefore, it can be hypothesized that T_out might have a potential impact in Dv50 by having an impact on the drying kinetics. Conversely, the omission of feed flow rate, reflected by the M_ratio, could possibly be attributed to the narrow operating range evaluated, limited by solubility challenges and the need to keep feed rates low. The FPF model encompassed F_atom and M_ratio as predictor variables, with the M_ratio having the most significant coefficient. This model follows the same trend as the Dv50 model, where higher atomization leads to a superior FPF, often linked to the smaller particle size. Regarding the M_ratio it has a negative effect on the FPF, which allows to extrapolate that high ratios are expected to negatively impact the powder aerosolization. This model resulted in a and of 0.84 and 0.79, respectively, both indicative of a robust fit of the model to the experimental data. No second order interactions were incorporated in either of the models, meaning in this situation SD optimization can rely only on the direct manipulation of process parameters. In terms of multicollinearity, the VIF for the predictors of each model were all approximately 1, confirming that there is no correlation between the variables. Furthermore, ANOVA analysis corroborated the models' statistical relevance in explaining the variance in the output data, evident through having both the p-values <0.5 and F-Ratio> Fcrit.

To further analyze the models and ensure their accuracy, external validation was conducted. Two experiments were conducted within the established design space as intermediate random points, and one experiment was conducted outside of the previously studied conditions, with both the T_out and M_ratio being tested above their upper level of the DoE. The points inside the design space were correctly predicted by both models, with relative errors between predicted and actual results being under 10%. However, for the point outside the studied range, the Dv50 model struggled to accurately predict the particle size. This is due to the incongruence between the models expected size reduction with increased temperature and the reality of droplet drying where temperature does not significantly affects particle size distribution. On the other hand, the FPF model accurately predicted the output, maintaining a relative error below 10%. Therefore, the models are reliable, having good predictive power within the studied design space, as is usual with empirical models. Moreover, the FPF model's omission of the T_out renders it to be more approximate to the expectation than the Dv50 model.

Overall, the results enabled the development of empirical models for two CQAs of DPIs. The models enabled the identification of the actual CPPs for the investigated CQA's and established relationships between them. It was concluded that the T_out and F_atom are the CPPs that influence particle size, which is not in agreement with what is commonly observed, and that F_atom and M_ratio are the CPPs that affect the FPF. The models provide insight into the optimization of the SD process to finely tailor powder properties, particularly concerning API encapsulation with a 3FN. Further enhancements could encompass expanding the DoE, for instance developing a central composite design and broader factor range. This would add useful information to the model, enabling the investigation of potential second order interactions or curvatures in the output data, rendering the model more robust and applicable.

473a Influence of Packing Fraction on Powder Permeability: A New Characterization Method

Aurelien Neveu, Marco Lupo, Filip Francqui, Geoffroy Lumay

Many industrial applications, such as pharmaceutical, food, additive manufacturing processes, use powders. It is, therefore, very important to precisely characterize these materials to properly control and optimize the processes. Many are the factors affecting the macroscopic behavior of the powders: particle properties (e.g. size and shape distribution), cohesive forces, friction and interlocking between the particles, stress state of the powder in the process, interaction with the surrounding fluid. The influence of air is very important in many phenomena such as the powder discharge rate from a silo, dispersion of an inhalation mixture in dry powder inhalers, the fluidization regime in fluidized beds, die filling in pharmaceutical and powder metallurgy processes. For example, in die filling of tablet manufacturing, air needs to be rapidly removed in order to bring the particles closer. In fact, the denser the powder in the die, the better the quality of the final tablet (Vreeman & Sun, 2022). The capacity of a material to allow a fluid (e.g. air) to pass through its bulk is called permeability. This property depends on powder porosity since the denser the powder, the lower the capacity of air to percolate through the material. At contrary, the looser the powder, the bigger the space between the particles and the higher the permeability. Therefore, permeability strongly depends on the packing conditions. Because of the shear deformation induced to the powder by the process, particles can rearrange to reach either a more or less compacted state, which in turn affects the permeability. A new approach has been developed to study the effect of packing conditions on permeability of the powders. It consists in compacting the powder bed by means of taps and in evaluating the permeability of the powder bed at different packing fractions, i.e. at different number of taps. By doing so it is possible to correlate the typical packing conditions, occurring in different industrial processes, to the resulting powder permeability. Powders with different shapes and particle size distributions, which are commonly used in different processes, such as 7 different kinds of lactose powders used as carrier for pharmaceutical tablets, have been tested. The results show that there is a direct effect of packing fraction of the powders on permeability.

This new approach aims to be a valid tool to investigate and predict the powder behaviour in typical industrial processes, where the interaction between air and the particles can be a crucial phenomenon, and to provide some useful advice to enhance the process efficiency.

473b Applications of Dynamic Vapor Sorption in Characterizing Agrochemical Solids

Ipsita Mishra, Matthew Russell

Detailed solids characterization is crucial for optimization and troubleshooting the processes involving solids. An important property studied during characterization is the interaction of solids with the solvents present in the system or ambient humidity. This information guides decisions about processing, storage conditions, and handling techniques. Dynamic vapor sorption (DVS) is a useful technique to carry out such a characterization. In this technique, solvent molecules are used as a probe and sorption/desorption profiles are obtained. The nature of these profiles and/or the hysteresis between the two provide information about the type of interaction between the solid and the solvent under consideration. The interaction could result in absorption, surface adsorption, formation of hydrates or solvates, or other changes in the solid form. In this work, interaction between different agrochemical solids and water has been explored with the DVS instrument obtained from Surface Measurement System and the above-mentioned scenarios were identified. Additional characterization, such as particle size distribution, microscopy, and thermal analysis, were also carried out to better interpret the DVS results.

473c Powder Characterization Using an Annular Ring Shear Cell Under Temperature and Humidity Control

Riya Shinde, Zankrut Vyas, Amit Sen, Abhishek Shetty, Sonia M. Razavi, Gerardo Callegari, Fernando Muzzio

Powders exhibit various levels of complexity; hence determining their physical and chemical properties and categorizing them can help with powder handling and processing. Powder characterization plays a critical role in various industries, including pharmaceuticals and catalyst manufacturing. It helps to classify powders depending on their flow and bulk properties, such as flowability, compressibility, bulk density, etc. These categories, in turn, help identify surrogates, selection of process parameters, and any pre-treatments the powder must undergo before processing. Various methods and equipment have been designed for powder characterization. Recently an air-bearing rheometer equipped with a convection temperature device (CTD), a humidity generator and a powder shear cell was developed to characterize effects of intrinsic and extrinsic factors on powder flowability The above-mentioned system can perform shear tests with robustness and accuracy. The instrument is especially advantageous due to its sensitivity and ability to produce data at very low torque levels. Effects of temperature and humidity can be systematically studied with such a system. This allows us to predict powder behavior in systems with variable temperature/humidity conditions during processing. This
rheometer uses an annular ring geometry to measure the shear properties. The shear stress is calculated based on the torque measured by the rheometer head to which the shear cell geometry is connected. The rheometer with a dedicated software is able to control the system temperature, humidity and apply normal force and strains consistent with shear cell testing. A 2-factor 3-level design of experiments (DoE) was performed to investigate the effect of temperature and humidity on three powders (Do we need to mention what these powders are). The highest temperature was selected to be below the melting point of the powder. A model was constructed for predicting cohesion values within the range of temperature and humidity that was investigated.

473d Capillary Imbibition Techniques Used to Characterize the Structural Changes in Microcrystalline Cellulose Due to Wet and Dry Processes

Pablo Ravazzoli, Maria Balbi, Nazareth Ceschan, Fernando Muzzio, Gerardo Callegari, German Drazer

Microcrystalline cellulose (MCC) is a common excipient used in the pharmaceutical industry. It has been reported that wetting and subsequent drying decreases its binding ability during tabletting as well as its swelling (and contribution to disintegration) due to irreversible hydrogen bonding, otherwise known as 'hornification'. Hornification of MCC could result from intermediate manufacturing processes such as wet granulation and it is therefore important to understand its effect on the physical properties of the powder and its producs, e. g. pharmaceutical tablets. Here, we consider how wetting and drying cycles changes the wettability properties of microcrystalline cellulose (MCC-Avicel 101). In order to characterize wettability, we perform the following experiments: (i) drop penetration in a slightly compressed power bed, (ii) capillary imbibition in a packed powder column, and (iii) drop penetration into tablets. We consider three cases for the MCC powder, the original dry powder as well as the powder obtained after one or two cycles of wetting and drying. In each case, the powder is sieved and separated into two samples, depending on particle size, $25 < \phi < 75$ μm and B) $75 < \phi < 125$ μm.
The drop penetration experiments are performed with both water and a silicon oil (Polydimethylsiloxane-PDMS) drops, and an effective contact angle between water and MCC is obtained from the comparison assuming the silicon oil perfectly wets the powder. The penetration of water is a combined effect between capillary penetration and swelling, and the effective contact angle is a result of these effects. We investigate the effect of hornification due to wetting and drying cycles on this effective contact angle. The column imbibition experiments are also performed using water and PDMS and we mesure the liquid mass uptake as function of time to obtain the effective contact angle from the comparison. Finally, we study the penetration times of water drops deposited on tablets made at 8 kN compression force (for one of the particle size range $25 < \phi < 125$ μm). The penetration of water in tablets obtained by direct compression of MCC show the (undesired) effect of hornification, in that penetration time increases.
We compare the effective contact angle between water and MCC obtained with the three different characterization methods for the original MCC as well as MCC after one and two wetting and drying cycles. We present the effect of hornification for both powder samples consisting of larger and smaller powder particles and the correlation between the different methods. Finally, we discuss the possibility to determine relevant effective contact angles and their transferability in different processes for powder materials that swell.

473e Stress Distribution and Relaxation in a Powder Column Under Compression

Amalia Thomas, Jamie Clayton

This work investigates the distribution of stresses transmitted by a compressed powder from a piston acting vertically on a powder column, onto the containing cylindrical vessel base and sidewalls. For this purpose, we used a pressure-sensitive Fujifilm Prescale film and a thin flexible SingleTact capacitive sensor to visualise pressure distributions and to quantify stresses over areas as small as 50 mm^2. The latter was the smallest and most sensitive flexible force sensor commercially available at the time this work was carried out, and it proved to have sufficient resolution to differentiate between the compaction properties of different powders. Of the materials tested, Whole Milk Powder and Corn Flour had the highest compressibility as measured using an FT4 Powder Rheometer, but in this study we measured significant differences in their stress distribution and retention behaviour in comparison to the less compressible Eskal (calcium carbonate) samples.

From the footprint left by the compressed powder on the pressure-sensitive film on the vessel base, we differentiated between two forms of stress distribution, according to the powder column height. For short columns the pressure transmitted through the powder onto the base was uneven. In fact, the shorter the distance between piston and base, the stronger and more localised we observed pressure to be transmitted by the powder on the base. In contrast, for powder columns taller than a critical value, the pressure on the base appeared evenly distributed (except around the outer edge of the base where the pressure is lower), albeit decreasing in magnitude the farther the base is from the piston.

In the region immediately below the piston we measured very high radial stresses on the vessel sidewall, in some cases even higher than the pressure applied by the piston. This result validates similar observations of a radial stress distribution along the compressed surface made through different methods (ball indentation, neutron diffraction), that the authors were inconclusive about at the time. We conjecture that the magnitude of the maximum radial stress in this area is powder dependent, as is the rate at which it decreases with depth from the piston. In Corn Flour, for example, the radial stress near the outer edge of the piston was particularly high and it decreased rapidly with depth. In this case there was a large radial stress gradient with depth and high shear between consecutive powder layers. We expect that powders that exhibit such high stress gradient are more susceptible to defects such as capping and lamination, so we propose that measurements of radial stresses in this narrow region may support a useful method for the prediction of tabletting defects. The SingleTact sensor proved to be sensitive enough and to have high enough resolution to differentiate between the stress gradients of the powders evaluated for this study, and future studies will focus on the differentiation of the compaction properties of pharmaceutical powders. As wider-range and more sensitive technology becomes available, we will also aim to reduce the dimensions of these experiments, to study powder behaviour at a scale more representative of real industrial processes.

The measurements obtained for this work were also highly differentiating in the powder stress retention properties. We measured that for all powders tested the fraction of the stress retained after compression increased linearly with depth. But the absolute fraction and its gradient of increase with depth are largely powder-dependent. We corroborated that powders that are less permeable and whose particles are more frictional and that tend to interlock (as measured from Permeability and Specific Energy measurements, respectively, using an FT4 Powder Rheometer), relax significantly less after compression. The results presented in this work may have a significant impact in the characterisation and design of formulations for optimal tabletting performance.

473f Characterization of Pharmaceutical Powder Flowability Using a Powder Flow Cell Coupled to an Air-Bearing Rheometer

Amit Sen, Zankrut Vyas, Riya Shinde, Abhishek Shetty, Gerardo Callegari, Sonia M. Razavi, Fernando Muzzio

The change in flow properties of powders at different stages of processing can greatly impact the quality of the final product. To ensure consistent outcomes, it is crucial to effectively characterize and comprehend the flow properties at each stage. Therefore, various powder characterization methods have been developed in literature and one needs to pick the one which mimics the state of powder in the process of interest to get the most meaningful information [1].

The focus of our work centers on a powder flow cell accessory mounted on an air-bearing rheometer which has a diverse array of testing methods [2]. In particular, Warren Spring Cohesion (WSC), Cohesion Strength (CS), and Dynamic Flow tests were explored on this device. The instrument is especially advantageous due to its sensitivity and ability to produce data at very low torque levels since it's on an air-bearing platform. The WSC evaluates the flowability of a powder under a defined consolidated state (applied normal stress). The flowability of the powder is assessed by measuring the maximum torque registered by a submerged vane geometry, which is also the failure point of the sample and is called the WSC strength. Prior to the shear to failure step the penetration of the vaned geometry is also monitored (normal force evolution) which gives an indication of the penetration work for the said consolidated sample.

The CS test involves a powder preparation step which includes fluidizing the powder to erase any sample memory (loading effects) on the measurement. The Cohesion strength test describes the internal resistance of the powder to flow, and thereby is a measure of powder flowability under no load. This CS method can only be used with powders that can be fluidized. Comparability of different powders is gauged by probing similar stress states in the powder bed, which is ensured by measuring at the same position in the powder bed (i.e., using the same volume/mass). Data is normalized with aerated or loose density values when very dissimilar materials (having very different densities) are compared.

The dynamic flow cell testing allows the calculation of various dynamic flow parameters for quality control, such as basic flowability energy (BFE), stability index (SI), flow rate index (FRI) and specific energy (SE).The dynamic flow cell testing provides vast sensitivity in terms of determining flowability in both confined and unconfined states.

More than 10 pharmaceutical powders were tested using the above-mentioned tests. It was found that the WSC method is more applicable to materials that exhibit more of a cohesive behavior. When the pre-compaction of the powder bed was increased, materials that were categorized as cohesive demonstrated higher WSC values. The CS method finds its applicability in pharmaceutical manufacturing operations such as fluidized bed processes. Specific Energy, measured in the Dynamic flow test, showed much higher values for very cohesive powders.

In conclusion, these characterization studies must be analyzed in tandem to understand the flowability of Pharmaceutical Powders used in different manufacturing stages. By gathering all the data obtained from these techniques in an effective way, extensive material databases can be generated which are not prevalent in the industry. Using the powder flow cell coupled to an air-bearing rheometer provides users with useful information, while preserving standards of easy-to-use, precision, and repeatability.

[1] Schulze, Dietmar. 'Powders and bulk solids.' *Behaviour, characterization, storage and flow. Springer*22 (2008).

[2] Hassanpour, Ali, Colin Hare, and Massih Pasha, eds. *Powder flow: Theory, characterisation and application*. Royal Society of Chemistry, 2019.

333a Semi-Mechanistic Prediction of Residence Time Metrics and Mixing Dynamics in Single-Screw Extrusion Via a 2-D Convection-Diffusion Model Combined with Machine Learning

Ashley Dan, Erik Holmen Olofsson, Shashank V. Muddu, Ninna H. Jokil, Jesper H. Hattel, Rohit Ramachandran

In an extrusion process, mixing dynamics is an important factor that correlates with the quality of extrudates, such as viscosity. The degree of mixing is related to the distribution of materials in the extruder, thus affecting the resultant flow pattern and system residence time distribution (RTD). Generally, RTD measurements have been used to quantify axial mixing, as it provides the mean residence time (MRT) and can be used to estimate the amount of dispersive mixing [1]. A greater MRT within the extruder barrel promotes efficient mixing, which affects the viscosity of the extrudates in the output. RTDs are influenced by process parameters (screw speed, feed rate, temperature profiles), equipment parameters (screw design), and material properties (feed viscosity) [2].

A starved-fed extruder, as compared to a flooded extruder, is highly applicable and industrially relevant as it eliminates issues related to bridging and funneling in the feed hopper or slippage on the barrels in the extruders [3]. In a starved extruder, another important metric that influences RTDs is the fill length, which represents the length of the extruder where barrels are fully filled with materials. Olofsson et al. [4] developed an advanced CFD model simulating starved extruders, which predicted residence time based on the fill length. These models were validated using experimental data of pressure drop observed at the extruder outlet, and the RTD curves were also compared to the analytical solutions. However, these models are computationally expensive, taking approximately 48 hours to obtain a single RTD curve. Therefore, there is a need to develop a computationally efficient model that can be used to predict RTDs and mixing dynamics

For this work, a hybrid model is developed, whereby a ML component is integrated with a 2-D convection-diffusion model, to accurately predict RTDs and mixing dynamics. Here, mixing dynamics are represented by the relative standard deviation (RSD) across the cross-section of the extruder. A ML-based model will be trained as a surrogate to replicate the CFD simulations, providing information on velocity fluxes across discretized radial and axial grids along the extruder, which are required inputs to the convection-diffusion model. CFD simulated data corresponding to a 3x3x3 factorial design DOE with independent parameters, including screw speed, feed rate, and feed viscosity, will be generated to train and validate a model that can account for these variations as model inputs. The dispersion coefficient will act as a tuning parameter in the model and will be estimated through model calibration minimizing errors between the predicted and CFD simulated RTD and RSD results. Finally, the model validation will be carried out using the estimated dispersion coefficient to test the prediction accuracy of the model outputs. The proposed semi-empirical model, involving data-driven predictions of velocity fluxes trained using data from first principles CFD models, coupled with a 2-D convection-diffusion model, will be able to predict RTDs and axial RSDs as a function of process parameters and feed properties.

1. Kumar, A., *Experimental and model-based analysis of twin-screw wet granulation in pharmaceutical processes*. 2015, Ghent University.
2. Yeh, A.-I. and Y.-M. Jaw, *Modeling residence time distributions for single screw extrusion process*. Journal of Food Engineering, 1998. **35**(2): p. 211-232.
3. Giles Jr, H.F., E.M. Mount III, and J.R. Wagner Jr, *Extrusion: the definitive processing guide*

and handbook. 2004: William Andrew.
4. Olofsson, E.H., Roland, M., Spangenberg, J., Jokil, N.H., Hattel, J.H., A CFD-model with free surface tracking: Predicting fill level and residence time in a starve-fed single screw extruder. The International Journal of Advanced Manufacturing Technology.

333b Development of High-Speed Computing Method for Powder Mixing Using Machine Learning with Random Model

Naoki Kishida, Hideya Nakamura, Shuji Ohsaki, Satoru Watano

Powder mixing is ubiquitous in a wide variety of manufacturing sectors including chemical industries, and it plays an important role. The degree of mixing of powders (a measure of uniformity) has a significant effect on the quality of the final product obtained by processing the powders, and thus methods to predict the degree of mixing are required. Computer simulation using the Discrete Element Method (DEM) is a powerful tool for the analysis of powder mixing process. In a DEM, dynamics (position and velocity) of an individual particles are numerically solved. These calculations are performed at extremely small time-step (micro seconds order). Therefore, a large number of iterations are required to calculate the entire powder mixing process, and it is difficult to predict a practical mixing in an actual process.

Recently, some studies on the combination of DEM and machine learning (ML) have been conducted. In these studies, macroscopic bulk powder phenomena (angle of repose, kinetic energy, mixing index etc.) were predicted. However, there is no study on prediction method of an individual particle flow and mixing behavior. In this study, we developed an original machine learning model, namely recurrent neural network with stochastically calculated random motion (RNNSR), which allows long time scale powder mixing simulation with low computational cost and high accuracy. RNNSR learns individual particle dynamics from short-time DEM simulation results and predicts a powder mixing for a long time. RNNSR combines recurrent neural network with stochastic model to predict convective and diffusive mixing. RNNSR is evaluated in terms of degree of powder mixing, particle velocity, granular temperature, and computing speed.

333c High Order Mass Preserving Finite Volume Approach for Rennet-Induced Coagulation Equation

Mehakpreet Singh

The coagulation of casein micelles caused by milk-clotting enzymes is a critical step in the dairy industry for cheese manufacture. During enzymatic coagulation of milk, three distinct but overlapping processes occur (a) enzymic proteolysis, (b) coagulation, and (c) gelation. This study presents a first ever numerical approach based on finite volume scheme for a enzyme-induced coagulation of casein micelles (first two phases). The finite volume scheme is mainly concerned with adhering to the mass preservation and developed on the assumption that the particles are concentrated on the mean of the cell. The key advantages of the new technique are its simple mathematical formulation, robust enough to be implemented on any type of grid and can be tailored for different coagulation kernels. Due to the non-availability of the exact (analytical) results of number density functions, the accuracy of the new approach is tested and verified against the newly derived analytical results for numerous gelling and non-gelling coagulation kernels. The comparison demonstrates that the new approach predicts the results with high precision and closely matches with the exact results. In order to analyse the convergence behaviour of different order moments, various refined non-uniform grids have been taken into consideration.

333d Super Learner Based Recurrent Neural Network Modelling Twin-Screw Granulation

James D. Litster, Kai Liu, Mahdi Mahfouf, Daniel Coca

Purpose:

Continuous manufacturing has gained significant attention in recent years due to the advantages it offers over traditional batch manufacturing in the pharmaceutical industry. In pharmaceutical manufacturing process, several processes such as mixing, drying, granulation, milling and coating are employed. Granulation, specifically, is an important process in pharmaceutical formulation, excipient active pharmaceutical ingredient (API) coming with powder are formulated to the granules. Twin-screw granulation (TSG) is one such granulation technique that has become increasingly popular in the production of solid oral dosage forms. TSG is a complex process that involves the mixing, wetting, and granulation of powders in a twin screw extruder. The process is influenced by many factors such as screw configuration, screw speed, feed rate, feed material property and process conditions, which makes it difficult to control and optimize using traditional methods.

Recurrent neural networks (RNNs) are a type of neural network that can handle sequential data and capture the dependencies between variables over time, compared to the traditional feedforward neural networks, RNNs have cycle connection in the architecture that allow them to take into account the time series data. In TSG, the variables can change rapidly and interact with each other in complex ways, by modelling the temporal dynamics of the process, RNNs can provide more accurate predictions of the granule qualities, such as the particle size.

Methods:

Twin screw granulation simulations were carried out with Siemens gPROMS Formulated Product, in which the feed flowrate, liquid flowrate, screw configuration and feed solid material size are altered. In order to enhance variety and dynamics in the TSG simulation, the pseudorandom binary sequences (PRBS) signals were utilized for the feed and liquid flowrates within the condition ranges, meanwhile, different numbers of kneading elements were employed to vary the screw configurations while two distinct material sizes were used.

Two recurrent neural networks, namely the Layer Recurrent Neural Network (LRNN) and the Nonlinear Autoregressive with Exogenous Inputs Neural Network (NARXNN), were employed in this study. Both networks contained one input layer, one hidden layer, and one output layer. In the LRNN, the hidden layers are connected in a feedback loop, thereby enabling the network to retain a memory of the previous inputs and computations. On the other hand, the NARXNN's output at each time step depends not only on the inputs at the current time step but also on the inputs and outputs from previous time steps. To obtain the optimal prediction results, the number of nodes in the hidden layer and the time delay steps had to be identified for both RNNs. Figure 1 illustrates the diagrams of the two recurrent neural networks.

To improve the predictive accuracy of the two RNNs, the Super Learner algorithm was employed. This algorithm is a machine learning technique that uses an ensemble of multiple base models to create a more accurate and resilient predictive model. The algorithm employs an ensemble learning strategy in conjunction with cross-validation to combine the predictions of multiple base models, resulting in a final prediction that surpasses the individual model predictions in terms of accuracy.

Finally, the developed super-learner based RNN model is utilized to predict the particles size and porosity in twin-screw granulation. To evaluate the prediction accuracy of the model, a comparison is made between predicted values with the gPROMS simulated values, and the model accuracy is shown by root-mean-squared error (RMSE), mean absolute percentage error (MAPE) and coefficient of determination .

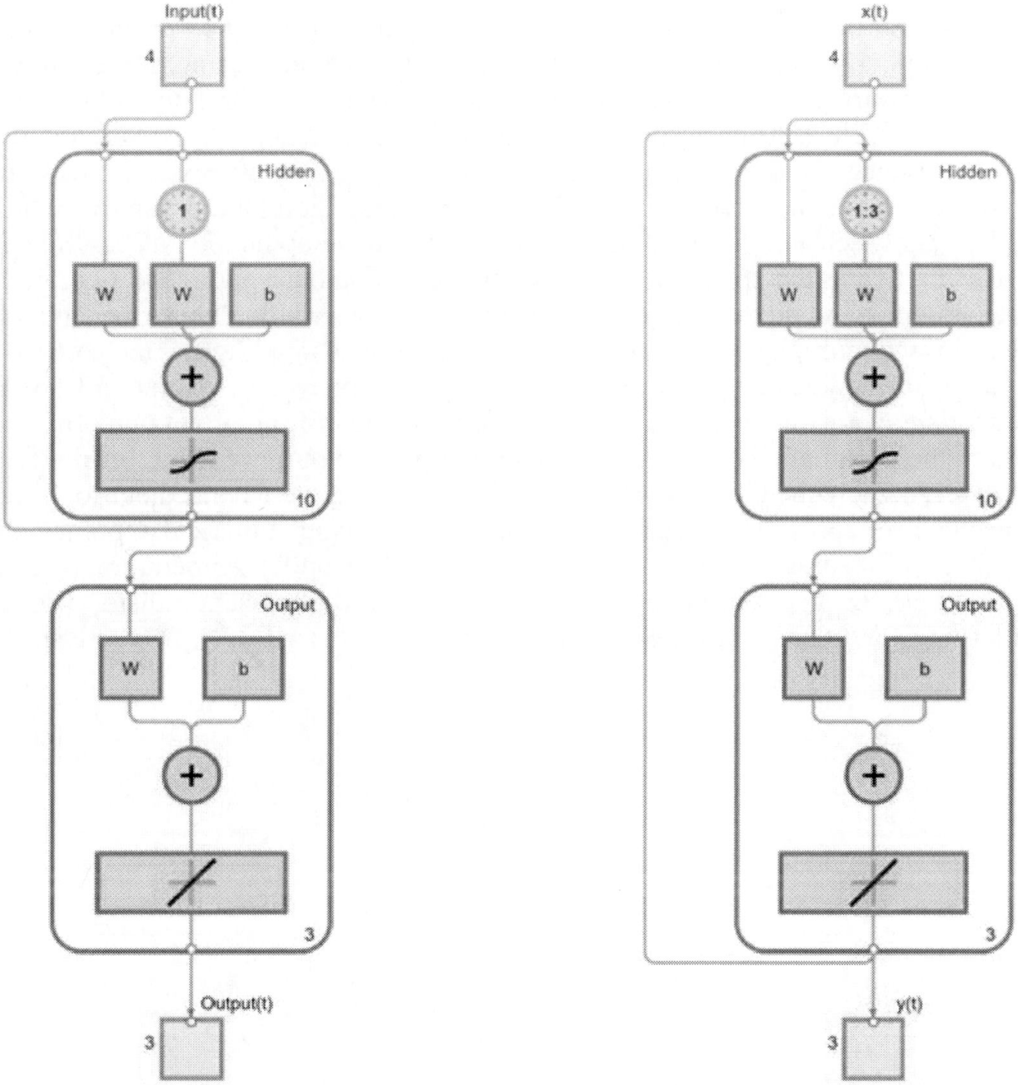

a

b

333e Predicting Drying Performance of Pharmaceutical Spray Dryers By Means of Multiphysics Simulations

Florian Ries, Ute Lander, Mike Weinberg, Ali Guerbuez, Andreas Thomann, Regina Reul, Andreas Gryczke

Spray drying is a process of forming a dry powder out of liquid solution by drying with a hot gas flow. It is often applied in the pharmaceutical industry in the manufacturing of thermally sensitive materials including pure active pharmaceutical ingredients (APIs), amorphous solid dispersion (ASDs) of purely soluble APIs or of formulations of pharmaceutical proteins. Despite of many years of industrial experience, the development of successful spray drying processes is not straightforward and often requires time- and material consuming experiments. Therefore, predictive modeling approaches are highly relevant for a fast-track and low-risk development of successful spray drying processes. This motivates the present work, in which we evaluate the applicability of multiphysics simulation technique to predict the drying performance of industrial scale pharmaceutical spray dryers. The proposed multiphysics simulation approach consists of a Lagrangian particle tracking (LPT) within a wall-modeled large eddy simulation (LES) framework. Accurate inflow droplet statistics for the spray are obtained from an embedded volume of fluid direct numerical simulation (VOF-DNS) of the pneumatic nozzle atomizer. For comparison purpose, experiments on an industrial scale spray dryer, namely the GEA PHARMA-SD Type PSD-1, were carried out for pure acetone and excipients/acetone solutions. It is shown that the proposed holistic numerical approach allows accurate predictions of spray drying process conditions such as outlet temperature, residence time distribution, particle size distribution. Finally, the effects of turbulent heat transfer, fluid dynamics, mixing and drying conditions on the performance of the GEA PHARMA-SD type PSD-1 spray dryer are discussed.

333f Design Space Optimization of a Sachet Filling Device Using Discrete Element Method (DEM)

Amir Esteghamatian, Matthew Santangelo, Hong-Guann Lee, Rolf Larsen, Jordan Arcangel, Diogo Gomes Lopes, Mario Schaefer, Daniel O. Blackwood, Pankaj Doshi

A pharmaceutical sachet filling device was modelled with the Discrete Element Method (DEM) using a cohesive contact model. The parameters for the contact model were calibrated by reproducing experimental flowability and compression data. The device comprises a horizontal screw feeder that periodically feeds a product hopper to keep the hold-up mass constant. A vertical auger draws the material from the product hopper into sachets at fixed doses, and a diamond-shaped agitator continuously mixes the material inside the product hopper. Various process parameters allow the independent control of the hold-up mass in the product hopper, the discharge rate of the material, and the mixing intensity in the product hopper. A DEM model was developed to optimize the design space, identify areas of risk in terms of powder segregation, and provide guidance for de-risking strategies. A special focus was placed on the impact of process parameters on mixing quality, particle velocities, and powder bed shape. The results indicate a high mixing quality across a wide range of operating parameters. The product hopper with the mixing agitator dampens the upstream concentration fluctuations—an effect that is investigated using a Residence Time Distribution (RTD) model. Less desirable operating points were identified by measuring the variability in mass and content at various locations within the device and in the sachets. Finally, a modification to one device part was suggested to improve the mass flow quality in the horizontal screw feeder section and further improve the overall mixing performance of the device.

333g A Hybrid Multivariate Model for Understanding Fluid Bed Wet Granulation Process Parameters Impact on Tablet Physicals

Sydney Butikofer, Maitraye Sen, Salvador García-Muñoz

Throughout the drug product development phase, models are developed and utilized to demonstrate process understanding. Systemic models provide the benefit of combining multiple unit operations, which allows the researcher to explore multivariate interactions across the production train. However, many complex models are time consuming to construct and require copious amount of data to generate high prediction accuracy. This work describes the development of a simple and easy to implement multivariate model of a drug product manufacturing process which utilizes fluid bed wet granulation followed by continuous direct compression.

Fluid bed wet granulation is a complex process as it involves several phenomena such as fluidization, granulation, and drying occurring simultaneously. To bypass mechanistic description of complex dynamic relationships between process parameters and responses, a partial-least-squares model was developed from the experimental data that was used to predict granule particle size. Subsequently, the granule particle size was used as an intermediate variable to relate fluid bed wet granulation process parameters to the resulting tablet physicals (e.g., tablet tensile strength). Combined, these relationships generate an integrated multivariate model which uses fluid bed wet granulation process parameters and compression stress as inputs to predict tablet tensile strength.

333h Utilising Science-Based Compaction Models for Drug Product Development: Tech Transfer and Flowsheet Simulations

Yash Melkeri, Indu Muthancheri, Stefan Bellinghausen, Lok Hang Wong, Mario Rousselin-Morales, Yasser Jangjou, Abhishek Paul, Nicolas Sanchez, Robert Yule, Ananya Chowdhury, Dana Barrasso

Roller and tablet compaction are key process steps in the production of pharmaceutical tablets. Because experimental data is limited or costly at early stages of development, an efficient process development workflow is key to make efficient use of the data available. A range of modelling solutions are available to understand the process behaviour and predict suitable process conditions for the different stages of development. However, the selection of the most appropriate modelling and experimental methodology for a given problem remains a challenge in pharmaceutical process development. Additionally, flowsheet solutions are useful to investigate the effect on downstream process and determine the desired conditions of a production line. The overall objectives of employing these solutions are to design the production process with minimal experimental effort and to reduce process variability to meet product specifications.

In this work, we will present an efficient design workflow for compaction tech transfer from compaction simulator to production equipment. This design workflow is based on dimensionless numbers and science-based models that are calibrated using targeted experimentation, taking into account the typical data available at different stages of pharmaceutical development. A flowsheet model is constructed based on the calibrated models of the relevant process steps. This flowsheet model is applied for the holistic process design.

230a Digital Manufacturing of Lithium-Ion Battery Electrodes Via Discrete Element Method (DEM)

Ruihuan Ge, Denis Cumming, Rachel Smith

Lithium-ion batteries are now widely used in the energy storage field, and most of the materials used to manufacture lithium-ion battery electrodes are particulates in form. In this work, Discrete Element Method (DEM) simulations are used to investigate the manufacturing process for lithium-ion battery electrodes such as dry powder deposition and calendering processes. For the calendering process, electrode mechanical and structural properties were investigated by using DEM. Detailed stress and microstructural evolutions at microscopic particle scale were analyzed. Inspired by the bimodal pore size distribution of electrode structures, a random field method is proposed to generate the multiple phase porous carbon binder domain (CBD) structures. By using advanced numerical approach, a process-microstructure-performance linkage is established. The results presented here demonstrate that DEM provides a highly promising way to digitalise the manufacturing process of the next generation battery electrodes.

230b Supported-Amine Adsorbent Generation Via Particle Molecular Layer Deposition for Direct Air Capture of CO_2

Hailey Loehde-Woolard, Bergen Evans, Kent J. Warren, Alan Weimer

As the adverse impacts of anthropogenic climate change continue to become more prevalent and intense, adaptation and mitigation technologies are becoming vital. Among the mitigation technologies, direct air capture (DAC) of CO_2 addresses emissions from non-stationary sources and could aid in addressing the existing atmospheric concentration. Further implementation of DAC technology in conjunction with an array of other fields that use CO_2 such as solar thermal chemical processing or reverse water gas shift reactions could produce liquid fuels and other desirable products. However, DAC faces complications due to the high processing volume of air and the low concentration of CO_2. Improvements in the CO_2 sorbents are needed to reduce cost and politically motivate DAC implementation. Traditional CO_2 capture technology relied on liquid amines which require substantial energy, and therefore substantial cost, to regenerate. To lower sorbent regeneration cost, solid adsorbent materials have been a focus of research for the past few decades. In particular, supported amines are attractive for DAC as they can operate under ambient conditions, regenerate at relatively low temperatures, and are tolerant of humidity. The current techniques of functionalization through liquid methods face limitations; grafting tends to have lower adsorption capacities while impregnation faces poor regeneration stability (Unveren 2017). We aim to introduce a novel technique to generate supported amine adsorbents with high adsorption capacity and stable regeneration.

Here, we employ particle molecular layer deposition (MLD) as a vapor-phase functionalization method. MLD is closely related to atomic layer deposition (ALD), involving a series of stepwise reactions between gaseous reactants on a solid surface to create thin films (Weimer 2019). We use MLD to deposit a covalently bonded aminopropylsiloxane network on particle supports. Two MLD chemistries have been employed: (3-aminopropyl)triethoxysilane (mono-amine) and N1-(3-trimethoxysilylpropyl)diethylene triamine (tri-amine). Sorbent materials were generated on varying support materials with varying amine composition based on a design of experiments. For comparison, materials were also generated through traditional liquid phase grafting. Characterization of the sorbents included BET surface area analysis, LECO elemental analysis, CO_2 chemisorption, and thermogravimetric analysis (TGA). The adsorption capacity was shown to increase as the number of MLD cycles increased for both chemistries. To understand the efficacy of the MLD functionalization method, we took the adsorption capacity on the basis of surface area. For example, the 25 cycle tri-amine on fumed silica MLD sample achieved 0.005 $mmol/m^2$ which is competitive with other surface functionalization methods and well performing adsorbents in literature. Ambient adsorption and low regeneration temperatures, below 100°C, were demonstrated by the mono-amine and the tri-amine materials. These materials appeared robust over cycling conditions between 25°C and 80°C and appeared stable for adsorption/desorption cycling temperatures below the deposition temperature of 150°C. This work demonstrates a new functionalization method and capture material to the field of solid CO_2 capture research that could aid in widespread CO_2 DAC implementation.

Ünveren, E. E.; Monkul, B. Ö.; Sarıoğlan, Ş.; Karademir, N.; Alper, E. Solid amine sorbents for CO 2 capture by chemical adsorption: A review. Petroleum 2017, 3 (1), 37

Weimer, A.W. Particle atomic layer deposition. J. Nanopart. Res. 2019, 21:9

230c Process Design for Carbon Black Size and Morphology

Georgios Kelesidis, Simon Benz, Sotiris E. Pratsinis

The structure (size and morphology) of carbon black (CB) largely determines its performance in tires, inks, batteries etc. (Kelesidis & Pratsinis, 2021). Here, the impact of CB process synthesis variables, i.e. temperature, T, precursor (i.e. acetylene) flow rate, Q, and volume fraction, φ, on CB structure (size and morphology) is elucidated by discrete element modeling (DEM; Kelesidis et al., 2017) and compared to experimental data in the literature. The CB structure is quantified by the mobility, d_m, and primary particle, d_p, diameters, effective density, ρ_{eff}, and fractal dimension, D_f. Decreasing Q or T enhances the CB yield, surface growth and agglomeration. This increases the mean d_m and d_p, consistent with experimental data from laminar flow ethylene pyrolysis reactors. The CB made at high Q or T is less compact than that made at low Q and T, having up to 50% smaller ρ_{eff} and 30% smaller D_f, in excellent agreement with experimental data also. Increasing the φ (e.g. by increasing the precursor flow rate), results in CB aggregates (covalently-bonded CB spherules) and increases the mean d_m and d_p, explaining commercial furnace black data. These CB aggregates are quite compact having up to 45% larger D_f and larger ρ_{eff} than those of ethylene black. So process design diagrams (Fig. 1) generated by the present DEM simulations can be used to guide design and optimization of CB for various applications (e.g. batteries, Kelesidis et al., 2022) from first principles.

References:

Kelesidis, G. A., and Pratsinis, S. E. 2021 A perspective on gas-phase synthesis of nanomaterials: Process design, impact and outlook. Chem. Eng. J. 421:129884, doi.org/10.1016/j.cej.2021.129884.

Kelesidis, G.A., Goudeli, E., and Pratsinis, S.E. 2017. Morphology and mobility diameter of carbonaceous aerosols during agglomeration and surface growth. Carbon 121:527–535, doi.org/10.1016/j.carbon. 2017.06.004.

Kelesidis, G. A., Rossi, N., and Pratsinis, S. E. 2022. Porosity and crystallinity dynamics of carbon black during internal and surface oxidation. Carbon 197:334-340, doi.org/10.1016/j.carbon.2022.06.020.

Figure 1. Design diagram of CB structure by varying the volume fraction, T and residence time.

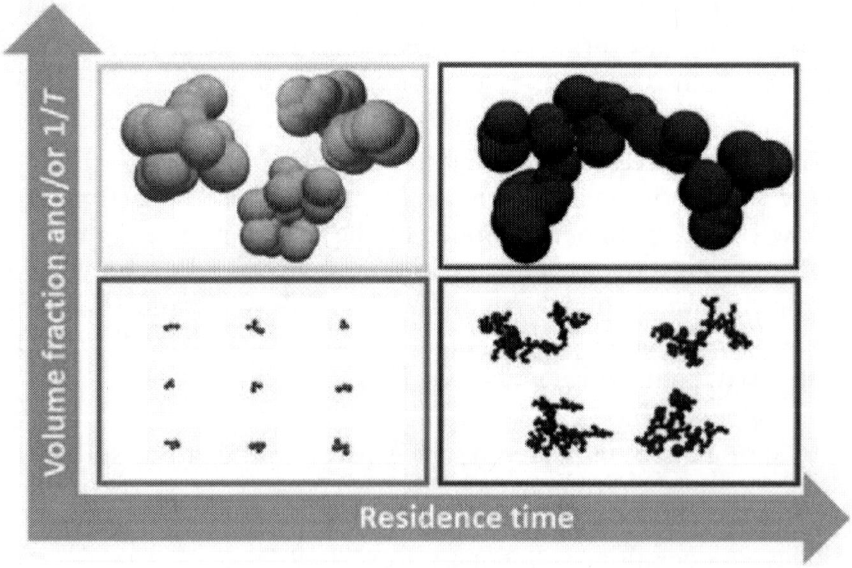

230d Aerosol-Based 3D Printing to Improve Resolution and Simplicity

Shalinee Kavadiya, Pratim Biswas

3D printing additive manufacturing processes have made prototyping and manufacturing easier and quicker due to their cost-effectiveness, automation, and capacity to produce complex structures. There are numerous 3D printing techniques currently available, such as fused deposition modeling, powder bed fusion, stereo-lithography, and others. Essential characteristics to consider for a 3D printing process are – resolution (capability to print small dimensions), versatility (capability to print a wide variety of materials and is easy to use), and speed (single step and high throughput). The existing techniques limit in at least one of these characteristics. Fused deposition modeling, for instance, is easy to operate but can only print polymers and has a maximum resolution of 100 μm, which is also hard to achieve. Powder bed fusion, a common technique for 3D printing metals, is complex and requires a high-power laser to melt the particles and a special enclosure to create a stable particle bed for printing, but it still lacks high resolution (< 50 μm). Moreover, stereo-lithography has good resolution but can only print photocurable polymers.

A simple aerosol-based 3D printing (A3DP) process that can print a wide range of materials at small-length scales (< 50 μm) and without requiring any special environment during printing will be discussed. While there are some aerosol-based printing methods (e.g., aerosol jet printing), our A3DP technique prints particles that are melted in the printing nozzle (and not on the surface). The A3DP method is a particle-based 3D printing method to create the structure. In this process, the feed material is aerosolized and introduced into the printer. The aerosolized droplets are dried, generating solid particles. The particles melt inside the nozzle that translates on the printing surface and prints the aerosols. The aerosol-based 3D printing takes advantage of the continuous aerosol process and combines it with 3D printing to achieve high-resolution continuous printing with precise control over the properties of the deposited structure. Examples of printed 3D structures with polyethylene glycol with a resolution of 41 μm will be described. Additionally, the feedstock can be controlled at any time to obtain the desired resolution and material. The ability to generate size and composition-controlled particles and their subsequent printing makes it easy to 3D print small dimensions. Currently, we are designing the printing nozzle by simulating the particle printing in COMSOL Multiphysics and developing sync between the particle generation and printing process to print structures with high precision and resolution.

The talk will discuss the role of several important process parameters, such as nozzle temperature, printing surface temperature, and nozzle-to-printing surface distance. This impact phenomena such as fluid flow, particle size, particle velocity, melting of materials, melting and solidification rate, eventually affecting the print resolution and quality. This simple, continuous, and high-resolution 3D printing technology can help print complex structures requiring high-resolution, such as microdevices for biomedical, military, and space applications.

Reference

Kavadiya and Biswas (2023), Improving the resolution and simplicity through aerosol-based 3D printing, in preparation

230e Swellable Cross-Linked Biopolymer to Enhance Ink Rheology and Mechanical Properties of 3D Printed Thin Film Incorporated with Pharmaceutical Active

Farzana Rony, Brian Rogers, Destinee Clay, Asmaa Alawbali, Shamsuddin Ilias, Mohammad Azad

Oral drug delivery constitutes 52% of the marketed drug product (i.e. tablet, capsule, etc.) [1]. However, swallowing a pill is a bottleneck for about 28% of patients predominantly children and geriatrics [1]. Hence, a polymer-based oral thin film is a promising alternative oral delivery route for pharmaceutical actives or drugs [2]. 3D Printing is an advanced manufacturing technology fast adopted in the pharmaceutical industry to make personalized medicine by tuning the dosage [3]. This research assesses the suitability of sodium starch glycolate (SSG), a swellable cross-linked biopolymer primarily used as a dispersant, to enhance the rheology of printing ink and the mechanical properties of the 3D printed thin film incorporated with Fenofibrate (BCS Class II) a cholesterol-reducing drug. 30 g batch of water-based printing ink (in the form of a paste) was prepared. The materials compositions of the ink include sodium alginate (base biopolymer), Fenofibrate (20%, w/w), SSG (property enhancer), glycerine (plasticizer), and polyvinylpyrrolidone (solubilizer or wetting agent). To obtain a uniform paste, Thinky mixture ARE-310 was used for 10 minutes at 2000 rpm. Ink rheology was evaluated using flow/viscosity, amplitude sweeps (1Hz frequency with a shear strain range from 0.001 to 1000%), and thixotropy tests. The thixotropy test comprised three intervals- the first and third intervals contained oscillation and the second interval was a rotation to deform the ink. Pressure-assisted microsyringe (PAM) type 3D printer (Cellink BioX) with printing conditions (410 mm diameter nozzle, an infill density of 50%, pressure of 60 kPa, and speed of 10 mm/s) were utilized to print these thin films (10 mm x 15 mm x 0.35 mm). The 3D-printed film was dried in an oven at 40°C for 24-48 hours. The viscosity curves show good shear thinning properties - the viscosity gradually decreases with the increase of shear rate which is an important property for film 3D printing. The thixotropy test shows a fast recovery which requires further investigation to correlate with film 3D printing. The films show good mechanical properties and testing results (Young's modulus, tensile strength, and elongation at break) are within the range found in the literature. Overall, it was a successful 3D-printed film. Other physical characteristics of the film such as dissolution test, scanning electron microscopy (SEM), etc. are in progress. Differential scanning calorimetry (DSC) and X-ray diffraction (XRD) will be done to ensure no drug degradation is observed. The findings will be discussed.

References:

[1] Borges, A. F., Silva, C., Coelho, J. F., & Simões, S. (2015). Oral films: Current status and future perspectives II—Intellectual property, technologies and market needs. Journal of Controlled Release, 206, 108-121.

[2] Elbadawi, M., Nikjoo, D., Gustafsson, T., Gaisford, S., & Basit, A. W. (2021). Pressure-assisted microsyringe 3D printing of oral films based on pullulan and hydroxypropyl methylcellulose. International Journal of Pharmaceutics, 595, 120197.

[3] Azad, M. A., Olawuni, D., Kimbell, G., Badruddoza, A. Z. M., Hossain, M. S., & Sultana, T. (2020). Polymers for extrusion-based 3D printing of pharmaceuticals: A holistic materials–process perspective. Pharmaceutics, 12(2), 124.

230f Combined Effects of Temperature and Moisture on Polymer Powders

Filip Francqui, Aurelien Neveu, Marco Lupo, Geoffroy Lumay

Polymer powders are found in numerous industrial applications. Especially, with the emergence of powder bed-based additive manufacturing, polymers are one of the most used materials in processes such as selective laser sintering (SLS) or binder jetting. In both cases, thin layers of powder have to be successively deposited on the powder bed, with the requirement to produce flat and homogeneous layers. The spreadability of the powder is thus an essential parameter for these applications and directly depends on the powder properties.

The cohesion is an important property of fine powders and is usually the main driving parameter for flowability. Cohesive forces are mostly due to the presence of capillary bridges between the particles and/or the setting of electrostatic charges arising from tribocharging mechanisms. Exposure of the powder to higher relative humidity will lead to modifications of the moisture content that directly influence the strength of the capillary bridges. Also, electrostatic properties will be influenced by the moisture content, the higher the moisture content the lower the charge buildup inside the powder. Indeed, the presence of the capillary bridges helps to set a low resistivity network that helps to dissipate the charges through the material. Therefore, two cohesive regimes are linked to the moisture: at low moisture content, the cohesion is driven by electrostatic, and at high moisture content the cohesion is driven by capillary forces.

In addition, temperature elevation leads to a modification of the surface properties of the particles, contributing to a change in surface friction as well as the global elasticity of the grains. Also, the moisture content will decrease when the powder is exposed to a higher temperature. Therefore, there are combined effects of temperature and moisture on cohesion that cannot be decoupled.

In this study, the effect of temperature and moisture has been investigated for different polymer powders. The moisture content has been varied by preconditioning the samples in a controlled atmosphere for a wide range of relative humidity. Influence on powder behavior is assessed with an improved tapped density analysis allowing to measure accurately the evolution of bulk density after each tap (GranuPack, Granutools, Belgium). The packing analysis is complemented by a flowability analysis performed in a rotating drum (GranuDrum, Granutools, Belgium). To evaluate the effect of temperature a special version of the GranuPack, the GranuPack High Temperature, allowing to heat the powder up to 200°C during the measurement has been used. The results gathered during this study allow a better understanding of the combined effects of temperature and moisture on polymer powders behavior.

230g Surface Growth of Soot By Acetylene Pyrolysis Using Reactive Molecular Dynamics

Anindya Ganguly, Xiaoke Wang, Georgios Kelesidis, Somesh Roy, Eirini Goudeli

Carbonaceous particles, or soot particles, are emitted from combustion sources, such as gasoline and diesel engines and coal-powered furnaces and power plants. Despite the prevalence of carbonaceous particles in our environment, in power generation and even in material production, there are still large gaps in our understanding of their formation and properties. The soot formation process has benefited from much more focus on soot precursors and polycyclic aromatic hydrocarbons (PAHs) formed in combustion, yet chemical reactions are still being discovered. The subsequent steps in soot formation, leading to large PAHs and particle growth via surface growth and agglomeration are far less understood, even though they determine the volume fraction of soot emissions. This creates barriers in developing an accurate representation of soot particles in climate models, leading to the underestimation of climate forcing (Ramanathan & Carmichael, 2008). The precise growth rate of soot nuclei till the formation of large aggregates is typically based on assumptions regarding the particle morphology (Mathieu et al., 2009) and chemical composition (Mosbach et al., 2009) or adjustments to better match experiments.

Here full-atom reactive molecular dynamics (MD) simulations are employed to capture the dynamics of incipient soot formation and growth by acetylene pyrolysis. The growth of the soot nanoparticles is monitored to derive a precise reaction rate constant accounting for soot cluster morphology, aromaticity, and composition during its growth. Figure 1 shows the temporal evolution of the volume-equivalent diameter, d_v (left axis), and the molecular weight, M_W (right axis), of the largest carbonaceous cluster formed at each timestep at 1650 K. Insets show snapshots of the largest cluster as a function time, indicating the formation an amorphous shell consisting of aliphatic chains and a core consisting of aliphatic and aromatic ring structures. Furthermore, the effect of process conditions (e.g., temperature, particle number concentration) on soot dynamics is quantified. The MD-obtained temperature-dependent reaction rate constant is interfaced with a monodisperse particle dynamics model (Kelesidis & Kholghy, 2021) to predict the soot volume fraction, finding excellent agreement with measurements in premixed ethylene flames.

MD simulations are used to track the growth of soot clusters formed by acetylene pyrolysis. The effect of temperature on the structure, composition, and morphology of soot as well as on its growth rate is quantified. The robust MD-obtained reaction rate derived here can be interfaced with a monodisperse particle model to elucidate the growth of soot in fractal-aggregates and thus simulate accurately the evolution of the soot volume fraction.

Mathieu, O., Djebaïli-Chaumeix, N., Paillard, C. E., & Douce, F. (2009). *Combustion and Flame*, 156(8), 1576-1586.

Mosbach, S., Celnik, M. S., Raj, A., Kraft, M., Zhang, H. R., Kubo, S., & Kim, K. O. (2009). *Combustion and Flame*, 156(6), 1156-1165.

Kelesidis, G. A., & Kholghy, M. R. (2021). *Materials*, 14(14), 3882.

Ramanathan, V., & Carmichael, G. (2008). *Nature geoscience*, 1(4), 221-227.

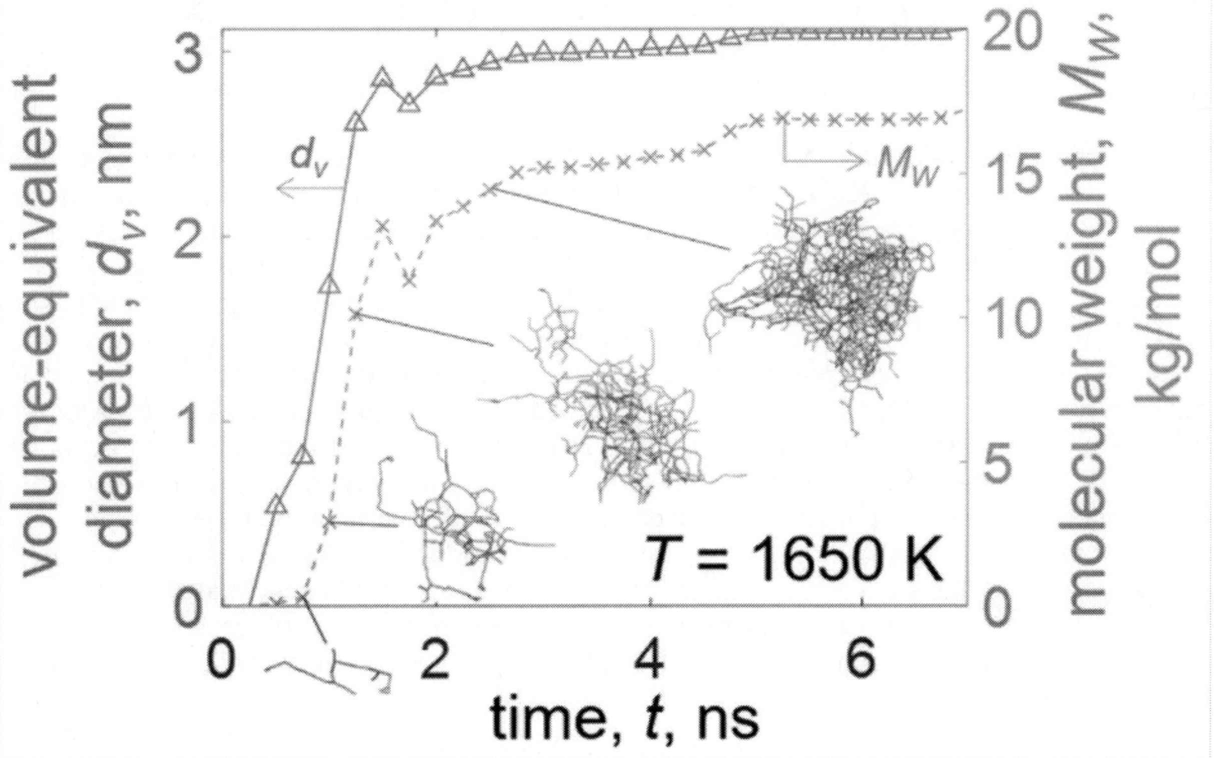

Figure 1. Temporal evolution of the volume-equivalent diameter, d_v (left axis), and the molecular weight, M_W (right axis), of the largest carbonaceous cluster formed at 1650 K by acetylene pyrolysis. Insets show the evolution of soot structure (black lines correspond to C atoms; H atoms are omitted).

492a Compositing Sulfur with Conductive-Additive Particles for All-Solid-State Lithium Sulfur Batteries By Using Continuous Hot-Melt Kneading Process

Hideya Nakamura, Motoshi Iwao, Hiromi Miyamoto, Eiji Hayakawa, Shuji Ohsaki, Satoru Watano

All-solid-state lithium-sulfur (ASS-Li/S) batteries have recently attracted considerable attention owing to their high energy density and safety. To produce the cathode of an ASS-Li/S battery, sulfur (the cathode active material) must be combined with a conductive additive (electron conductor) because of the electronic insulating property of sulfur. Therefore, a compositing process of sulfur and conductive additives is necessary to produce ASS-Li/S batteries. However, existing compositing methods are neither scalable nor productive. Herein, for the first time, a hot-melt kneading process, as a scalable and productive compositing process, to produce composite particles of sulfur and a conductive additive for ASS-Li/S batteries is employed. The composite particles prepared from the hot-melt kneading process show larger particle sizes, less fine conductive additive particles, and better flowability than the simple mixture. The obtained composite particles have a matrix-type structure, in which conductive additive particles exist even inside the sulfur particles. Concerning the electrochemical performance, compositing sulfur and a conductive additive using the hot-melt kneading process improves the electrochemical performance because of the matrix-type structure of the composite particles. Moreover, by estimating the productivity of the process, the study demonstrates that the hot-melt kneading process has a significantly better productivity compared with conventional compositing processes.

Reference

[1] M. Iwao, H. Miyamoto, H. Nakamura, E. Hayakawa, S. Ohsaki, S. Watano, Continuous Compositing Process of Sulfur/Conductive-Additive Composite Particles for All-Solid-State Lithium Sulfur Batteries, Adv. Energy Sustain. Res. (2023). doi:10.1002/aesr.202200206.

492b A Review of the Influence of Particle Size and Surface Chemistry on the Electrochemical Performances of Cathode, Anode and Electrolyte Materials

Guangliang Liu, Wenjing Li, Yingmeng Zhang, Lei Yao, Haipeng Yang

With batteries gradually being the essential power sources for consumer electronic devices, portable power tools, electric vehicles, and so on, the demand for high performance batteries with higher energy density, quicker charging time and better safety performance, becomes urgent. There are many particle processing technologies involved in battery manufacturing, such as particle size reduction, particle surface modification, powder mixing, et al. , which knowingly impact final battery performance. Here, the significance of the influence of particle size and surface chemistry on the electrochemical performances of cathode, anode and electrolyte materials in literatures is reviewed.

492c Dry Mixing and Its Impact on Powder Properties for Dry Coating of Lithium-Ion Battery Electrodes

Marcella Horst, Arno Kwade

In the context of climate change and the expansion of electromobility, the demand for lithium-ion batteries (LIB) is increasing rapidly. Electrode and cell production is associated with high energy consumption of approx. 40 kWh per kWh of battery capacity produced. As one of the largest energy consumers and cost drivers in the common electrode production, the drying process of the wet-coated electrodes offers a high potential for improving the economic efficiency, sustainability, and carbon footprint of the battery manufacturing process. A solvent-free processing of battery electrodes via a dry coating process represents a promising method to achieve these goals.

In this regard, the dry mixing of the powdered electrode components at the beginning of the process chain have a crucial impact on the performance of the dry coated battery electrodes. Within this process step, it is important to structure the electrode materials to achieve a well-tuned powder flowability being important for the following process steps (powder metering and film formation). While for wet film coating the binder is pre-dissolved in the solvent during slurry production, in the dry mixing process the binder must be homogeneously distributed in the powder mixture and the conductive additives must be well dispersed within the powder mixture. Furthermore, the binder together with the mixed powder must be able to build a network (e.g. by PTFE fibrillation) during the film formation while maintaining a sufficient electrical conductivity of the resulting electrode.

The impact of the mixing process in a high-intensity mixer on powder mixtures consisting of cathode-active material, carbon conductive additives and binder (PTFE), was investigated. Process-structure-property relationships were identified through various powder characterization methods (flowability coefficient, dynamic angle of repose, electrical resistance, Hausner ratio, particle size distribution). Figure 1A shows particle size distributions as a function of the specific mixing energy (here corresp. mixing time). With increasing specific mixing energy, the particle size distribution becomes narrower, shifts towards smaller particle sizes, and gets monomodal for high specific mixing energies. This represents the fibrillation, which is greatest at low mixing energies and the fibrils are broken up with increasing time. The fibrillation process is also presented in the photos in Figure 2, showing that after the premixing of active material and carbon additive, the addition of binder causes fibrillation. A special post-treatment leads to a more homogeneous product. The increasing flowability of the mixture during the mixing process consequently increases its bulk density. Its electrical resistance decreases due to improved particle contacts (see figure 1B). The fibrillated powders were also evaluated for their functionality in a dry coating process with powder metering and film formation via a multi-roll calender.

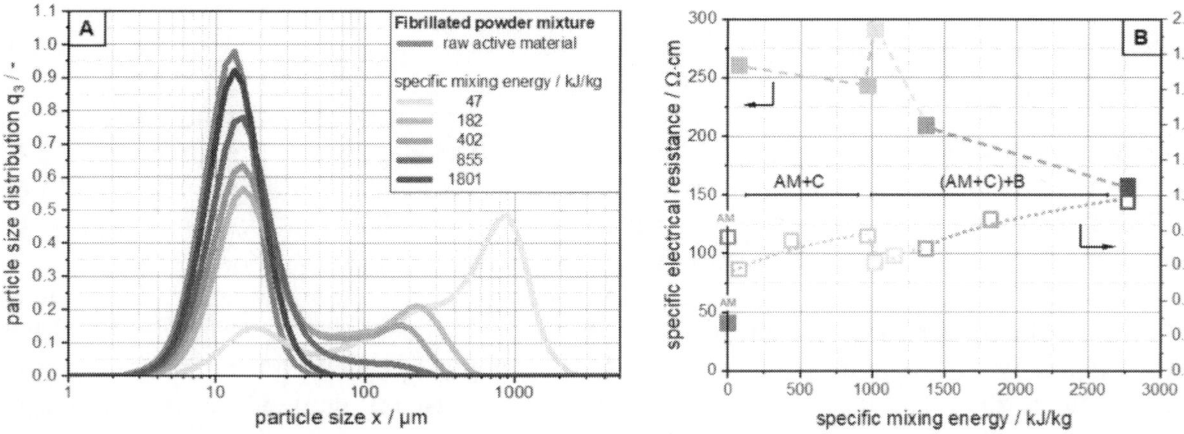

Figure 1: (A) particle size distributions and (B) bulk density and specific electrical resistance of a fibrillated powder mixture

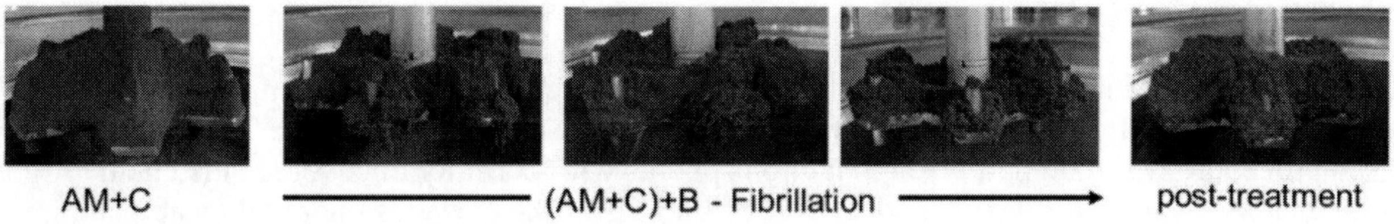

Figure 2: Fibrillation process of a battery cathode powder mixture

492d Predicting Fuel Cell Ink Aggregation and Drying

Harsh Srivastav, Clayton Radke, Adam Weber

Polymer-electrolyte fuel cells (PEFCs) hold promise as a viable alternative to combustion engine for heavy duty vehicles and long-term energy storage. They have been sought-after in these fields for enabling minimal emissions technologies with high efficiencies and energy densities. Of the cost of the PEFC, the most expensive and complex part of the cell is the PEFC catalyst layer, which is key for enhancing the performance of the entire device. Numerous studies have been performed to investigate the correlation between the materials that comprise the catalyst layer and the performance of the cell, but the overarching relationship between the two remains undetermined. PEFC catalyst layers are typically constructed with a carbon supported catalyst particle (most commonly platinum) with an ionomer, often a perfluorosulfonic acid (PFSA) such as Nafion, which secures the carbon suspensions in the ink dispersions by preventing the particles from crashing out as well acting as a binder after the solvent has been evaporated off. To aid in the exploration and comparison of such complex systems, a few recent studies have looked at utilizing the role of the pH of the ink dispersion.[1]

In this talk, we outline an approach to predicting the interactions between the particles through mathematical modeling using both macroscopic and microscopic ink properties. We utilize a kinetics-based procedure based on pairwise interaction spheres with partially uncovered side chain groups that simultaneously solves for the pH and sizes distributions of the particles under adjustable parameters, such as solvent and particle wt%. A novelty which has been included is the idea of backside repulsion of the particles, where the double layers on the opposite sides of interacting particles can further push them together, greatly impacting aggregation rates. Further, the outputs of this formulation will be used to ascertain fundamental properties of the catalyst layer such as the porosity using a space filling argument. To display the precision and robustness of this proposal, we juxtapose the predictions against experimental pH data for a range of concentrations and solvents. The intuition gained from this model will guide future design goals and process conditions.

Acknowledgements

This study was conducted under the Million Miles Fuel Cell Truck Consortium (M2FCT) funded by the Hydrogen and Fuel Cell Technologies Office in the Energy Efficiency and Renewable Energy Office of the U.S. Department of Energy under contract DE-AC02-05CH11231.

References

1. S. A. Berlinger, B. D. McCloskey, and A. Z. Weber, *J. Phys. Chem. B*, **122**, 7790–7796 (2018).

290a Creating a New Materials Circularity with Negative Emission Technologies

Ah-Hyung Alissa Park

With the increase in the global population and the growth of the world economy, global energy consumption has been increasing rapidly. Carbon capture, utilization and storage (CCUS) technologies are one of the approaches to decarbonize the power and industrial sectors, and create a new circular carbon economy. Among various options, mineral carbonation, which mimics the natural weathering of silicate minerals, has potential at scale relevant to climate change mitigation. As CO_2 *reacts with silicate minerals, carbon is stabilized in the form of insoluble solid carbonates for CCUS with permanence. In addition to natural minerals and mine tailings, a wide range of primary and secondary sources (e.g., alkaline industrial wastes (e.g., iron and steel slag), electronic wastes, waste PV and wind power systems etc) can be used as feedstock for carbon mineralization. Carbon mineralization using their own solid wastes and* CO_2 *leads to multifaceted environmental benefits including CCUS and solid waste management. Furthermore, these technologies can be employed to recover energy-relevant critical materials such as rare earth elements (REEs), Li, Co, Ni and Cu while sequestering carbon. The challenge is that silicate minerals and solid wastes are chemically complex and their dissolution kinetics are very slow and the recovery of metals from these feedstocks require multi-step separation technologies. In order to address these challenges and opportunities, we have focused on the fundamental understanding of dissolution and carbonation behaviors of unconventional resources and the integration of step-wise separations of energy-relevant critical materials creating a new materials circularity for our sustainable future.*

290b Chemical Looping Gasification and Reforming

Liang-Shih Fan

The concept of chemical looping reactions has been widely applied in chemical industries. Fundamental research on chemical looping reactions has also been applied to energy systems. Fossil fuel chemical looping applications were used with the steam-iron process for coal from the late 1800s to early1900s and were demonstrated at a pilot scale for synthetic natural gas production with the IGT HYGAS Process and the CO_2 Acceptor Process in the 1960s and 1970s. There are presently no chemical looping processes using carbonaceous fuels in commercial operation. Key technical factors that determine commercial viability of the technology lie in the sustainability of the reactivity and recyclability of the metal oxide oxygen carriers and the ability of configuring the reactor assembly for optimal operation and control of the solids flow system. The chemical looping system is developed on the circulating fluidized bed platform. The successful deployment of this technology requires thorough knowledge of two interconnected fields, i.e., metal oxide reaction engineering and particle science and technology. With now CO_2 emission control of great concern and process conversion efficiency enhancement of great interest, activities on research and development of chemical looping technology have resurfaced.

Specifically, chemical looping technology is a manifestation of the interplay among such key elements of metal oxide reaction engineering and particle science and technology as particle synthesis, reactivity and mechanical properties, flow stability and contact mechanics, gas-solid reaction engineering and particulates system engineering. This presentation will describe the fundamental and applied features of modern chemical looping technology in the context of the circulating fluidized bed platform that utilizes fossil and other carbonaceous feedstock. It will discuss the reaction chemistry, ionic diffusion mechanisms, metal oxide synthesis and thermodynamics, reactor design, and system engineering along with energy conversion efficiency and economics of the chemical looping processes for, specially, partial and selective oxidation for syngas and chemicals production. The Ohio States University has developed a number of advanced chemical looping gasification and reforming processes which have been licensed to industry for commercialization. This presentation will highlight the readiness of this technology commercialization.

290c Review of Particle Attrition and Attrition Test Methods

Jennifer Curtis

Particle attrition is a major concern in the petroleum industry. It occurs in many process units (e.g., fluid catalytic crackers, risers, standpipes, conveying lines, fluidized beds, and hoppers) that involve flowing particulates. Particle attrition results in a loss of valuable catalyst mass and can lead to degradation of product quality and performance, as well as environmental problems. Hence, understanding and mitigating catalyst attrition is very important for successful process unit operation, control and economics. While there have been many previous studies on particle attrition, it is a complex process which depends on many factors (particle hardness, particle size and shape, impact velocity, etc). This talk will discuss current experimental methods for assessing attrition and significant findings to date.

290d The Rheology of Pyroclastic Powders

Raffaella Ocone

Granular materials exhibit a broad range of intricate dynamic behaviours. The study of their hydrodynamics is extremely relevant in the chemical and process industries, where those materials are widely handled and produced. Understanding how internal (e.g., particles size and shape) and external (e.g., applied stresses, moisture content) physical properties impact on the flow behaviour of solid particles helps industrial practitioners handle and produce particulates in an efficient and less costly way. Building upon previous findings applied to the process industry, the talk explores the challenges associated with the dynamic behaviour of dry and wet granular material and discusses recent experimental and modelling efforts on the flowability of pyroclastic powders. Pyroclastic powders are investigated with the aim of predicting and managing the hazard resulting from volcaniclastic debris flows, natural phenomena which occur when a mixture of pyroclastic fallout/current deposits and water move down slopes under the action of gravity.

290e Bubble Self-Organization in Annular Gas-Solid Fluidized Beds

Kaiqiao Wu, Shuxian Jiang, Victor Francia, Marc-Olivier Coppens

Imposing periodic perturbations to granular systems by a fluctuating external force can generate self-organized patterns that are regular in time and space. An oscillating energy input in nonlinear, complex granular systems creates ordered structures due to energy dissipation. Drawing inspiration from granular media in nature, where patterns form on beaches and dunes due to the action of waves or gusts of wind, we apply the same mechanism that leads to regular pattern formation in complex systems to dynamically structure the bubble pattern in fluidized beds. In a pseudo-2D system, an oscillatory perturbation of the gas flow rearranges bubbles into hexagonal arrays with a controlled wavelength and bubble size, irrespective of bed width. The pattern's properties are determined by the gas flow's pulsation frequency, amplitude and offset. The flow structure is fully scalable and leads to much more controllable properties, such as bubble size distribution, residence time, and spatial distribution[1–3]. These unique features could allow us to bypass some of the challenges in conventional fluidized bed units, like flow maldistribution and non-uniform contact, as well as decouple conflicting design objectives, such as good solid mixing and gas-solid contact.

Facilitating the scale-up of such structured flows will allow engineers to design and operate a new class of fluidized beds, named dynamically structured fluidized beds. Recently, drawing from our fundamental investigations, we have been able to recreate a similarly structured flow in a cylindrical geometry, extending it from a flat-2D to an annular-3D configuration. This presents additional opportunities for scaling up gas-solid systems for, e.g., drying, coating, and reactions. Nevertheless, different from a flat-2D bed, two critical differences arise in the annular-3D system: it has no lateral walls and introduces curved boundaries. These may result in instabilities affecting the pattern formation process. In this contribution, we used both experiments and simulations to investigate the reproducibility of the 3D periodically structured pattern under different oscillatory flows. The bubble flows within the annular geometry exhibit adaptive self-organization, fitting along the circumference and showing some elasticity by slight rotation along the vertical axis. Compared with the flow properties in the flat-2D system, the experimental results demonstrate that the curved boundaries can stabilize the pattern, with minimal impact on the manifestation of the structured flows. This indicates the adaptability of these flow patterns across different configurations, which is very useful for particle processing and process intensification. Our results also demonstrate a multi-parametric operating window, which identifies a range of frequencies and amplitudes of the applied pulsations to produce structured flows in the annular-3D system.

Over many years, and through his extensive experience, Dr Ray Cocco has provided priceless advice, encouragement and technical insights that have allowed us to develop our ideas further to bridge the gap between fundamentals and practical applications, informing how nature-inspired chemical engineering could impact fluidized bed applications. We remain extremely grateful for this.

References

1. Wu, K.; de Martín, L.; Coppens, M.-O. Pattern formation in pulsed gas-solid fluidized beds–the role of granular solid mechanics. *Chem. Eng. J* **2017**, *329*, 4–14.
2. Francia, V.; Wu, K.; Coppens, M.-O. Dynamically structured fluidization: oscillating the gas flow and other opportunities to intensify gas-solid fluidized bed operation. *Chem. Eng. Process.* **2021**, *159*, 108143.
3. Francia, V.; Wu, K.; Coppens, M.-O. solids lateral mixing and compartmentalization in dynamically structured gas–solid fluidized beds. *Chem. Eng. J* **2022**, *430*, 133063.

Keywords: Pulsed fluidization, pattern formation, scale-up, CFD-DEM, bubble control

290f Democratizing CFD: Expanding the Use of Fluidization Modeling to Address Today's Greatest Challenges

James Parker, Jia Wei Chew, Peter Blaser, Ray Cocco

In tackling urgent sustainability challenges, the fluidized bed has emerged as a key technology, leveraging its proven effectiveness in diverse applications over the past 100 years. However, transitioning from lab-scale proof-of-concept to commercial-scale implementation is an expensive and lengthy process. Given the pressing nature of climate change-related processes, expediting the commercialization of these new concepts becomes all the more critical. Computational Fluid Dynamics (CFD) is an increasingly important tool used in fluidized bed technology development for reducing risks in scale-up and accelerating development. Unfortunately, the use of CFD, particularly in fluidization, remains largely confined to CFD practitioners and domain experts. This talk aims to explore the existing barriers preventing widespread use of CFD as a tool by all chemical engineers involved in technology design, scale-up, and optimization and what is being and needs to be done to circumvent these barriers.

112a Particle Technology Forum: 30 Years in Perspective

Liang-Shih Fan

In commemorating the 30 year anniversary of the formation of the AIChE Particle Technology Forum (PTF), I would like, as the Founding Chair (1992-1994), to share some reflections on the circumstances surrounding the genesis of this organization and recognize those pioneering researchers who shared the same vision for it and who devoted their time, energy and talent to realizing its inception. In celebrating the achievement signaled by this important milestone, I would like to offer my thoughts about this field and its topics of emphasis which are in the midst of transformation as well as the new and rising generation of engaged researchers. I will also discuss the efforts required to ensure the continued impact of the field on both academia and industry.

112d Influence of Mechanical Interlocking on the Flow Behavior of Granular Particles

Jennifer Curtis, Elizabeth Suehr

In order to obtain reliable predictions for the flow behavior of highly angular particles, such as Lunar and Martian soils, mechanical interlocking associated with sustained particle-particle interactions needs to be described. Mechanical interlocking gives rise to large particle-phase stress in granular soils and a significant resistance to deformation. In this study, the effect of particle interlocking in dilute and dense-phase granular flows is investigated via discrete element method (DEM) simulation of a 3D system of hook-shaped particles in simple shear flow using Lees-Edwards periodic boundary conditions. The effect of particle shape is assessed by progressively varying the degree of curvature of the hook-shaped particles. The particle shape varied from a straight, elongated particle to one in which the straight particle is folded in half. For each particle shape, a total normalized interlocking time and distribution of particle interlocks are determined over a range of solid concentrations. At low solid volume fractions, the normal and shear stresses are proportional to the inverse of the projected area of the particle in the plane perpendicular to the flow, consistent with the stress behavior of elongated particles [Guo et al., 2012]. For larger solid volume fractions, normal and shear stresses increases with increasing interlocking, showing a direct dependence between stress and degree of mechanical interlocking

112c How Standpipes Really Work

T. M. Knowlton, Surya Karri

Standpipes have traditionally been thought of as a device which allows solids to be transferred from a low pressure to a higher pressure. Standpipes can certainly do this, but their true function is not so limited. The actual purpose of a standpipe is to adjust the pressure drop across them so that the pressure drops around the circulation loop are in balance.

The four different types of standpipes and their operation will be reviewed. How each type of standpipe adjusts to balance the overall pressure drop in a circulation loop will be explained. Examples of standpipes adjusting to various operating conditions and solids flow rates will also be presented and described in detail.

112e Persistence in the Face of Adversity to Achieve your Goals

Alan Weimer

Excerpts will be taken from this open access paper published in Powder Technology (https://doi.org/10.1016/j.powtec.2023.118279) as personal examples for how to overcome professional barriers along the way as you pursue your endeavors. The path from laboratory curiosities to commercial powder processing plants had many obstacles to overcome. The manner in which this was done can be applied to everyday life - be it obtaining a M.S. or Ph.D., pursuing a career afterwards, or simply living your life to the fullest in a very challenging world. PTF provides for an incredible opportunity to network with all kinds of people and is by far the most academic/industry integrated organization that I am aware of. Many "forks in the road" opportunities are available from being an active participant in PTF; so, I encourage you to be an active participant in this fantastic organization.

112f Particle Technology in Tackling Global Challenges: *Where are the opportunities*

Jinghai Li

Particle technology is a crucial field in the science and engineering of this age. It is particularly relevant to tackling global challenges such as carbon neutrality, public health and so on. On the other hand, it is facing disciplinary challenges in the research paradigm shift from being reductionism and average-based to being interdisciplinary and structure-based, a shift that better fills in the knowledge gaps at different levels.

I am going to share with you what we have done over the past 40 years or so, covering the conceptual shift from multiscale to mesoscale, and then to mesoscience in the field of particle technology.

I will first introduce why the average approach should be replaced by a multiscale approach, and why mesoscales at different levels are so crucial. I will then propose, with evidences from different cases, that the general principle of compromise in competition (CIC) is likely to exist for all mesoscales, leading to the conclusion that the concept of mesoscience refers not only to "scale", but also to "regime".

Finally, I will give a mesoscale perspective for different disciplines and technological fields beyond its origin in particle technology. The presentation will conclude with the prospect of deducing simplicity and commonality from complexity and diversity. In essence, what we will probably achieve in mesoscience is crucial for more broadly tackling the global challenges and shifting the paradigm in science.

112g Happy 30th B'day PTF: A personal perspective on its international reach

Marc-Olivier Coppens

For 25 years I have had the pleasure to benefit from the PTF's community, grow with it, and serve it in various roles, while being in academic positions from TU Delft in the Netherlands (1998-2006) to Rensselaer in the USA (2006-2012) and UCL in the UK now, with particle technology related collaborations in other countries and continents. Through the lens of one of the AIChE Inter national Committee's founding members, I will discuss the importance of the PTF from an international perspective, as the PTF has always been uniquely welcoming and embracing international participation. It has been an inclusive home for particle technology's academic and industrial researchers and practitioners from around the world, at any stage of their career, especially through the events at the AIChE annual meeting with global participation. This has led to important exchanges that have shaped our field, but also social interactions that have given the PTF a human face. At 30 years young, the PTF is thriving as the global home for the particle technology community.

112i Particle Technology: Reflection from Canadian Activities

Jesse Zhu

This presentation will discuss some examples in particle science and engineering activities that are currently taking place in Canada and will reflect on the past, present, and future endeavors in the research and technology development in this field. Many research and development activities in Canada are directly associated with its vast nature resources, including tar sands, oil shale, and biomass with particle technology being their enabling technology for its commercial success. Academic and industrial interaction has been an important part of these efforts. Some examples surrounding the prolific interactions of researchers between Canadian organizations and the AIChE Particle Technology Forum on education and research in the field will be presented.

35a Advances in the Simulation of Dry Powder Inhalation Flows

Sankaran Sundaresan

The treatment of many respiratory illnesses relies on dry powder inhalers (DPIs) to deliver medication to patients' airways and lungs through inhalation. The active pharmaceutical ingredients (APIs) of the drug are particles with diameter $< 5\mu m$, that allow the air to carry them to the lungs. Due to their small size and cohesiveness, the API particles form bigger agglomerates, which adversely affect delivery to the lungs. To mitigate this problem, the API particles are usually mixed with coarser lactose carrier particles (50 to 100 μm), allowing the API particles to adhere to the carrier particles. An effective DPI design seeks to maximize the release of the API particles from the carrier particles and suppress their subsequent agglomeration or deposition on the DPI walls. In this presentation, I will illustrate the use of simulations to examine the fluidization, transport, and deagglomeration of the API-carrier agglomerates in a DPI device and the human airways.

Prof. Sankaran Sundaresan is the Norman John Sollenberger Professor in Engineering and Applied Science and Professor of Chemical and Biological Engineering at Princeton University, where he has worked since 1980. Prof. Sundaresan, known as 'Sundar', is a world expert on the modeling and characterization of multiphase flows. Sundar works on dispersed multiphase flows, frequently encountered in chemical reactors and separation devices, that manifest complex structures at different length and time scales – micro, meso, and macro scales, which influence the mixing, mass and heat transfer, and reaction processes. Over the years, Sundar has made important contributions to understanding how to simulate complex flow behavior and relate this information to engineering applications. Sundar won the Elsevier Lifetime Achievement Award, AIChE Particle Technology Forum, 2022, for his research and service. He is a friend and supporter of the particle technology field and we are looking forward to hearing about his work on fluidization, transport, and deagglomeration processes during Dry Powder Inhalation, which are critical for the effective delivery of drugs to lungs.

35b People and Tools Advancing the Science of Particle Technology

Richard Lueptow

Particle technology has many challenging aspects, but more tools than ever are available to tackle these problems including discrete element method simulations, combined fluid-particle simulation capabilities, advanced experimental methods like particle tracking velocimetry, continuum modeling developments, and advances in the understanding of the fundamental physics of particle interactions. The Particle Technology Forum is a remarkable gathering of sophisticated practitioners from a wide range of industries and talented academicians with a variety of advanced research skills and interests. The result is a group that is destined to continue to expand the tradition of the PTF in advancing cutting-edge high-quality research that is fundamentals-based and has meaningful impact on real problems in industry.

35c Ten years at PTF: Connect fundamental research with industrial scale problem solving

Yi Fan

In this talk, I will share my great experience with the Particle Technology Forum (PTF) of AIChE in the past decade and how PTF helped me transform from an academic researcher to an industrial practitioner in the chemical industry. I will present a few examples how the Dow Solids Processing R&D team utilized the process fundamentals and lab scale research to address full production scale challenges by collaborating with academic partners, manufacturing engineers, and even marketing specialists. The topics span from particle segregation in packaging and product design, gas-solids flow dynamics in cross-flow moving packed bed reactors, and solids flow in production-scale hoppers.

35e Progress in wet granulation process development - from a pharmaceutical industry application perspective

Maitraye Sen

Particle engineering is an intricate part of pharmaceutical drug product development processes involving powders. Designing for acceptable powder flowability and compactibility is often challenging and requires significant particle engineering. Although simpler routes like direct compression and dry granulation are preferred, wet granulation is often unavoidable due to the physical attributes of the powder formulation and the required drug loading in the dosage form. Owing to its wide-spread adoption, different types of wet granulation techniques have been studied extensively in literature.

This talk will review the traditional approaches towards developing wet granulation processes and present case studies on current approaches. The current approaches will highlight the faster and more innovative ways of conducting drug product process development and ensuring product quality, which fits the current business environment that emphasizes reducing the time to market in a cost-effective manner.

Bio – Maitraye Sen

Maitraye is a chemical engineer with experience in pharmaceutical process and product design, and expertise in process modeling & simulation. She completed her doctoral studies from Rutgers University, NJ, USA, followed by postdoctoral research at Purdue University, IN, USA prior to joining Eli Lilly and Company, IN, USA. Her work is on building streamlined process development workflows and control strategies by integrating different mathematical and experimental tools for speedy development of robust pharmaceutical manufacturing processes.

35f Spherical Agglomeration as a Tool for Manufacturing of Formulated Products: Mechanistic Understanding and Mathematical Modelling

Rachel Smith

The micro-scale interactions between particles and fluids are fundamental to a wide range of industrial manufacturing processes, including granulation, coating, drying, and fluidized bed processes, and it is these interactions and processes which drive the Particle Technology Forum (PTF) of the AIChE. Substantial understanding of the nano- and micro- scale interactions between particles and fluids has been generated, however the ability to use this knowledge to predict process performance, design manufacturing processes, and to tailor formulated particulate products to give desired performance is underdeveloped, and industry is often still reliant on substantial experimental trial and error approaches to product and process design. These challenges are key to the PTF, which continues to further our fundamental knowledge of particle processing, and improve our ability to apply this knowledge to address industrial problems.

Spherical agglomeration is a size enlargement process which has recently gained interest in the pharmaceutical industry. It addresses followability and compressibility related problems of irregular platelet or needle like crystals, by forming spherical-shaped agglomerates. This process is achieved by the addition of a binder liquid into a slurry of primary crystals which is often produced by anti-solvent crystallization. Currently, there is poor mechanistic understanding of the rate processes occurring for formation of spherical agglomerates, and there is a lack of useful tools to predict final agglomerate properties or process performance. In this presentation, recent spherical agglomeration research informed and inspired by work presented through the PFT will be discussed, and research spanning experimental observation, development of mechanistic process understanding, and computational modelling will be presented.

35d Particle Technology and Clean Energy Transition for Sustainable Future

Ah-Hyung Alissa Park

With the increase of the global population and the growth of the world economy and industrial sector, global energy consumption has also been increasing. Considering that the fossil fuels will still be one of the major energy sources for the foreseeable future, measures need to be taken to control the atmospheric CO_2 concentration. Carbon capture, utilization and storage (CCUS) technologies are one of the approaches to decarbonize the power and industrial sectors. Among various options, mineral carbonation, which mimics the natural weathering of silicate minerals, has a potential at scale relevant to climate change mitigation. As CO_2 reacts with silicate minerals, carbon is stabilized in the form of insoluble solid carbonates for permanent carbon storage or CO_2 utilization with permanence. If this reaction is carried out in an ex-situ reactor system, solid carbonates, high surface area silica and other minor components such as iron oxide can be produced with tailored properties and separated as value-added products. In addition to natural minerals and mine tailings, alkaline industrial wastes such as iron and steel slags can also be used as feedstock for carbon mineralization. Most of the industries producing alkaline solid wastes (e.g., steelmaking, cement and aluminum plants) are also point sources of anthropogenic CO_2. Thus, carbon mineralization using their own solid waste streams and CO_2 leads to multifaceted environmental benefits including CCUS and solid waste management. Furthermore, industrial wastes often contain other valuable components such as rare earth elements. The challenge is that silicate minerals and alkaline solid wastes are chemically complex and their dissolution kinetics are very slow. In order to address these challenges and opportunities, we have focused on the fundamental understanding of dissolution and carbonation behaviors of alkaline silicate materials and integration of step-wise separations of rare earth elements from these unconventional resources.

AUTHOR INDEX

AIChE
120 Wall Street, FL 23
New York, NY 10005-4020

ISBN 978-1-7138-9317-2

CURRAN ASSOCIATES INC.
proceedings
.com